Trends in Mathematics

Research Perspectives CRM Barcelona

Series Editors

Enric Ventura
Antoni Guillamon

Since 1984 the Centre de Recerca Matemàtica (CRM) has been organizing scientific events such as conferences or workshops which span a wide range of cutting-edge topics in mathematics and present outstanding new results. In the fall of 2012, the CRM decided to publish extended conference abstracts originating from scientific events hosted at the center. The aim of this initiative is to quickly communicate new achievements, contribute to a fluent update of the state of the art, and enhance the scientific benefit of the CRM meetings. The extended abstracts are published in the subseries Research Perspectives CRM Barcelona within the Trends in Mathematics series. Volumes in the subseries will include a collection of revised written versions of the communications, grouped by events.

For further volumes:
http://www.springer.com/series/4961

Extended Abstracts Fall 2012

Automorphisms of Free Groups

Juan González-Meneses
Martin Lustig
Enric Ventura
Editors

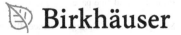

Editors
Juan González-Meneses
Universidad de Sevilla
Sevilla
Spain

Martin Lustig
Université Aix-Marseille III
LATP
Marseille
France

Enric Ventura
Universitat Politècnica de Catalunya
Fac. Matemàtica i Estadística
Barcelona
Spain

ISBN 978-3-319-05487-2   ISBN 978-3-319-05488-9 (eBook)
DOI 10.1007/978-3-319-05488-9
Springer Cham Heidelberg New York Dordrecht London

Library of Congress Control Number: 2014945551

2010 Mathematics Subject Classification: 20Exx, 20Fxx

Printed on acid-free paper

Springer is part of Springer Science+Business Media (www.springer.com)

Foreword

Since 1984 the Centre de Recerca Matemàtica (CRM Barcelona) has been organizing scientific events such as conferences and workshops which span a wide range of forefront topics in mathematics and witness the communication of new outstanding results. In the Fall of 2012, the CRM took the commitment to publish Extended Conference Abstracts (ECAs for short) emanated from scientific events hosted at the Centre, with the aim of quickly spreading new achievements, helping a fluent update of the state of the art and enhancing the scientific profit of CRM meetings. This initiative is hosted by Birkhäuser under the shelter of *Research Perspectives CRM Barcelona*, inside the established series *Trends in Mathematics*.

Books in this new series will contain collections of revised written versions of selected communications given by authors at the various conferences and workshops held at CRM, grouped by events. In coherence with the general purpose of the series, its mathematical texts, usually no longer than four pages, are thought to be more detailed than just conference abstracts, but less worked out than research papers in final form. In this sense, many of them will probably evolve into more detailed and specific papers published later on in regular research journals. The text of these ECAs can contain announcements of new results together with the necessary definitions and context, as well as possible motivation, open questions or conjectures; proofs are not required but short sketches are welcome. All the material will be scientifically revised (for coherence, contents, and quality) by organizers of the CRM events where each manuscript comes from. Finally, the CRM Editorial Board (and particularly the editors of the series) will take care of coordinating the process and editing the material. Special thanks are due to Carles Casacuberta, Editor-in-Chief of the CRM Editorial Board from 2008 to 2013, for his experienced support in structuring and editing this first volume of the series.

Barcelona, Spain

Toni Guillamon
Enric Ventura

Preface

From September to December 2012, a Research Programme on *Automorphisms of Free Groups: Algorithms, Geometry and Dynamics* took place at the Centre de Recerca Matemàtica (CRM), Bellaterra, Barcelona. It was coordinated by Juan González-Meneses (Universidad de Sevilla, Spain), Martin Lustig (Université Aix-Marseille III, France), Alexandra Pettet (University of British Columbia, Vancouver, Canada), and Enric Ventura (Universitat Politècnica de Catalunya, Barcelona, Catalonia).

During these intense four months, several scientific events took place, including an international conference, an advanced course, two workshops, and a weekly seminar, all of them with the active participation of many visitors invited to attend from several countries abroad. Of course, additionally, all participants had numerous occasions to host informal but fruitful conversations among themselves, discussing mathematical ideas which, in many cases, gave rise to new and interesting results, altogether in a very dynamic and productive research atmosphere.

In this first volume of the new subseries *Research Perspectives CRM Barcelona* published by Birkhäuser inside the series *Trends in Mathematics*, we present 17 Extended Abstracts corresponding to selected talks given by participants in the research programme. More than half of them came from the *Conference on Automorphisms of Free Groups* (held from November 12 to 16, 2012) and the rest came from talks given at the workshops or the weekly seminar held along the Fall of 2012. We hope that the presentation of this material under the present Extended Abstract form will help authors spread their recent research. Most of the short articles in this volume contain preliminary presentations of new results not yet published in regular research journals.

We would like to express our gratitude to CRM for hosting and supporting our research programme. We also convey our warm thanks to the CRM Director, Joaquim Bruna, and to the secretaries for providing great facilities and a very

pleasant working environment. Finally, thanks are due to all those who attended
the talks, for their interest, their active participation, and their enthusiasm towards
mathematics.

Sevilla, Spain   Juan González-Meneses
Marseille, France   Martin Lustig
Barcelona, Spain   Enric Ventura

Contents

Tits Alternatives for Graph Products

Yago Antolín

1 Graph Products

Let $\Gamma = (V, E)$ be a simplicial graph and suppose that $\mathfrak{G} = \{G_v \mid v \in V\}$ is a collection of groups (called *vertex groups*). The *graph product* $\Gamma\mathfrak{G}$ of this collection of groups with respect to Γ is the group obtained from the free product of the G_v for $v \in V$ by adding the relations

$$[g_v, g_u] = 1 \text{ for all } g_v \in G_v, g_u \in G_u \text{ such that } \{v, u\} \text{ is an edge of } \Gamma.$$

The graph product of groups is a natural group-theoretic construction generalizing free products (when Γ has no edges) and direct products (when Γ is a complete graph) of groups G_v with $v \in V$. Graph products were first introduced and studied by E. Green in her Ph.D. thesis [5].

Basic examples of graph products are *right angled Artin groups*, also called *graph groups* (when all vertex groups are infinite cyclic), and *right angled Coxeter groups* (when all vertex groups are cyclic of order 2).

2 Tits Alternatives

In 1972, J. Tits [10] proved that a finitely generated linear group either is virtually solvable or contains a copy of the free group $\mathbb{F}_2$ of rank 2. Nowadays such a dichotomy is called the *Tits alternative*.

Y. Antolín (✉)
School of Mathematics, University of Southampton, Highfield, Southampton, UK
e-mail: yago.anpi@gmail.com

J. González-Meneses et al. (eds.), *Extended Abstracts Fall 2012*,
Trends in Mathematics 1, DOI 10.1007/978-3-319-05488-9_1,
© Springer International Publishing Switzerland 2014

The Tits alternative can be naturally modified by substituting "finitely generated", "virtually solvable" and "contains a non-abelian free subgroup" with other conditions of a similar form. An example of a result of this type is a theorem of G. Noskov and E. Vinberg [8] claiming that every subgroup of a finitely generated Coxeter group is either virtually abelian or large. Recall that a group is said to be *large* if it has a finite index subgroup which maps onto a non-abelian free group. A large group always contains a non-abelian free subgroup, but not vice-versa.

Let us now formally describe the possible versions of the Tits alternative that we are going to consider. Let $\mathcal{I}$ be a collection of cardinals; a group G is said to be $\mathcal{I}$-*generated* if there is a generating set X of G and $\lambda \in \mathcal{I}$ such that $|X| \leq \lambda$.

Definition 1. Suppose that $\mathcal{I}$ is a collection of cardinals, $\mathcal{C}$ is a class of groups and G is a group. We say that G *satisfies the Tits alternative relative to* $(\mathcal{I}, \mathcal{C})$ if, for any $\mathcal{I}$-generated subgroup $H \leq G$, either $H \in \mathcal{C}$ or H contains a non-abelian free subgroup.

The group G *satisfies the strong Tits alternative relative to* $(\mathcal{I}, \mathcal{C})$ if for any $\mathcal{I}$-generated subgroup $H \leq G$ either $H \in \mathcal{C}$ or H is large.

Finally, the group G *satisfies the strongest Tits alternative relative to* $(\mathcal{I}, \mathcal{C})$ if for any $\mathcal{I}$-generated subgroup $H \leq G$ either $H \in \mathcal{C}$ or H maps onto a non-abelian free group.

For example, let $\mathcal{I}_\omega$ be the collection of all countable cardinals and $\mathcal{C}_{\text{vab}}$ be the class of virtually abelian groups. The result of G. Noskov and E. Vinberg [8] in this language becomes as follows: finitely generated Coxeter groups satisfy the strong Tits alternative relative to $(\mathcal{I}_\omega, \mathcal{C}_{\text{vab}})$.

Let $\mathcal{I}_f$ be the collection of all finite cardinals. Other families of groups known to satisfy the Tits alternative relative to $(\mathcal{I}_f, \mathcal{C}_{\text{vab}})$ are word hyperbolic groups [6], $\text{Out}(\mathbb{F}_n)$ [3], where $\mathbb{F}_n$ is the free group on rank n, and groups acting freely and properly on a finite dimensional CAT(0) cubical complex [9].

Basic examples of groups satisfying the strongest Tits alternative (relative to any collection of cardinals and the class of abelian groups) are free and free-abelian groups. Our Theorem 5 below provides the first non-trivial example of groups satisfying the strongest Tits alternative.

3   The Theorems

Our theorems state that, under very natural conditions, the various forms of Tits alternatives are stable under graph products.

We will be interested in collections of cardinals $\mathcal{I}$ and classes of groups $\mathcal{C}$ satisfying the following properties:

(P0) (*Closed under isomorphisms*) If A, B are groups, $A \in \mathcal{C}$ and $A \cong B$, then $B \in \mathcal{C}$.

(P1) (*Closed under $\mathcal{I}$-generated subgroups*) If $A \in \mathcal{C}$ and $B \leq A$ is an $\mathcal{I}$-generated subgroup, then $B \in \mathcal{C}$.

(P2) (*Closed under direct products of $\mathcal{I}$-generated groups*) If $A, B \in \mathcal{C}$ are $\mathcal{I}$-generated, then $A \times B \in \mathcal{C}$.

(P3) (*Contains the infinite cyclic group*) $\mathbb{Z} \in \mathcal{C}$.

(P4) (*Contains the infinite dihedral group*) If $\mathbb{Z}/2\mathbb{Z} \in \mathcal{C}$ then $\mathbb{Z}/2\mathbb{Z} * \mathbb{Z}/2\mathbb{Z} \in \mathcal{C}$.

Our first theorem is the following.

Theorem 2. *Let $\mathcal{I}$ be a collection of cardinals and let $\mathcal{C}$ be a class of groups enjoying the properties* (P0)–(P4). *Suppose that Γ is a finite graph and $\mathfrak{G} = \{G_v \mid v \in V\}$ is a collection of groups. Then the graph product $G = \Gamma\mathfrak{G}$ satisfies the Tits alternative relative to $(\mathcal{I}, \mathcal{C})$ provided each vertex group G_v with $v \in V$ satisfies this alternative.*

It is easy to see that all of these properties are necessary, since free products and direct products are special cases of graph products.

In order to state our theorem about the strong Tits alternative, we need an extra property.

(P5) (*$\mathcal{I}$-locally profi*) If $A \in \mathcal{C}$ is non-trivial and $\mathcal{I}$-generated then A possesses a proper finite index subgroup.

Theorem 3. *Let $\mathcal{I}$ be a collection of cardinals and let $\mathcal{C}$ be a class of groups enjoying the properties* (P0)–(P5) *such that $\mathcal{I}$ contains all finite cardinals or at least one infinite cardinal. Suppose that Γ is a finite graph and $\mathfrak{G} = \{G_v \mid v \in V\}$ is a collection of groups. Then the graph product $G = \Gamma\mathfrak{G}$ satisfies the strong Tits alternative relative to $(\mathcal{I}, \mathcal{C})$ provided each vertex group G_v with $v \in V$ satisfies this alternative.*

Examples of classes of groups with properties (P0)–(P5) for $\mathcal{I} = \mathcal{I}_f$ are the classes consisting of virtually abelian groups, virtually nilpotent groups, (virtually) polycyclic groups, (virtually) solvable groups and, more generally, elementary amenable groups.

To see that property (P5) is necessary for the claim of Theorem 3, consider the case where no group from $\mathcal{C}$ contains non-abelian free subgroups. Suppose that $A \in \mathcal{C}$ is a non-trivial group without proper finite index subgroups. Then $A * A$ will possess no non-trivial finite quotients. It follows that $A * A$ cannot be large; on the other hand, $A * A \notin \mathcal{C}$ as it contains a non-abelian free subgroup.

Finally, to state our theorem concerning the strongest Tits alternative, we need a definition and a new property.

Definition 4. For a collection of cardinals $\mathcal{I}$, we define a new collection of cardinals $\mathcal{I}'$ by saying that a cardinal λ belongs to $\mathcal{I}'$ if and only if $\lambda + 1 \in \mathcal{I}$.

(P6) (*$\mathcal{I}'$-locally indicable*) If $A \in \mathcal{C}$ is non-trivial and $\mathcal{I}'$-generated, then A has an infinite cyclic quotient.

Basic examples of groups satisfying (P6) with $\mathcal{I} = \mathcal{I}_f$ are torsion-free nilpotent groups. An extreme example of groups satisfying (P6) are torsion-free groups when $\mathcal{I} = \{0, 1, 2\}$.

Theorem 5. *Let $\mathcal{I}$ be a collection of cardinals and let $\mathcal{C}$ be a class of groups enjoying the properties (P0)–(P3) and (P6). Suppose that Γ is a finite graph and $\mathfrak{G} = \{G_v \mid v \in V\}$ is a collection of groups. Then the graph product $G = \Gamma\mathfrak{G}$ satisfies the strongest Tits alternative relative to $(\mathcal{I}, \mathcal{C})$ provided each vertex group G_v with $v \in V$ satisfies this alternative.*

It is easy to see that the assumption of (P6) in Theorem 5 is indeed necessary.

4 Some Corollaries

The previous theorems provide a variety of new examples of groups satisfying the different versions of the Tits alternatives. We focus only on corollaries of Theorem 5.

Corollary 6. *Let $\mathcal{C}$ be a non-empty class of torsion-free groups closed under isomorphisms, direct products and taking subgroups. Suppose that G is a graph product with all vertex groups from $\mathcal{C}$. Then for any 2-generated subgroup $H \leqslant G$, either $H \in \mathcal{C}$ or $H \cong \mathbb{F}_2$.*

This corollary generalizes a classical result of A. Baudisch [2], who proved that any two non-commuting elements of a right angled Artin group generate a copy of $\mathbb{F}_2$. In fact, we are now able to say much more about subgroups of right angled Artin groups. Indeed, by taking $\mathcal{I} = \mathcal{I}_\omega$ and $\mathcal{C}$ to be the class of finitely generated torsion-free abelian groups, and applying Theorem 5, we obtain the following result.

Corollary 7. *Any subgroup of a finitely generated right angled Artin group is either free abelian of finite rank or maps onto $\mathbb{F}_2$.*

A classical result of R. Lyndon and M. Schützenberger [7] states that if in a free group three elements a, b, c satisfy $a^m = b^n c^p$ for $m, n, p \geq 2$, then these elements pairwise commute. Combining this with the previous corollary we obtain the following result.

Corollary 8. *If three elements a, b, c of a right angled Artin group G satisfy $a^m = b^n c^p$ for $m, n, p \geq 2$, then these elements pairwise commute.*

This corollary generalizes a theorem of J. Crisp and B. Wiest [4, Theorem 7], who established this claim in the case when $m = n = p = 2$.

Acknowledgements This talk was based on the research paper [1], which is a joint work with Ashot Minasyan.

References

1. Y. Antolín, A. Minasyan, Tits alternatives for graph products (2011). arXiv:1111.2448
2. A. Baudisch, Subgroups of semifree groups. Acta Math. Acad. Sci. Hung. **38**(1–4), 19–28 (1981)

3. M. Bestvina, M. Feighn, M. Handel, The Tits alternative for Out($\mathbb{F}_n$), I. Dynamics of exponentially-growing automorphisms. Ann. Math. (2) **151**(2), 517–623 (2000)
4. J. Crisp, B. Wiest, Embeddings of graph braid and surface groups in right-angled Artin groups and braid groups. Algebr. Geom. Topol. **4**, 439–472 (2004)
5. E.R. Green, Graph products. Ph.D. thesis, University of Leeds, 1990
6. M. Gromov, Hyperbolic groups, in *Essays in Group Theory*. Mathematical Sciences Research Institute Publications, vol. 8 (Springer, New York, 1987), pp. 75–263
7. R.C. Lyndon, M.-P. Schützenberger, The equation $a^M = b^N c^P$ in a free group. Mich. Math. J. **9**, 289–298 (1962)
8. G.A. Noskov, E.B. Vinberg, Strong tits alternative for subgroups of coxeter groups. J. Lie Theory **12**(1), 259–264 (2002)
9. M. Sageev, D.T. Wise, The Tits alternative for CAT(0) cubical complexes. Bull. Lond. Math. Soc. **37**(5), 706–710 (2005)
10. J. Tits, Free subgroups in linear groups. J. Algebra **20**(2), 250–270 (1972)

Asymptotic Approximations of Finitely Generated Groups

Goulnara Arzhantseva

1 Introduction

The concept of approximation is ubiquitous in mathematics. A classical idea is to approximate objects of interest by ones simpler to investigate, and which have the required characteristics in order to reflect properties and behavior of the elusive objects one started with.

Looking for approximation in geometric group theory, first we adapt this fundamental approach. We discuss both its well-established appearance in *residual properties* of groups and its recent manifestation via *metric approximations* of groups such as sofic and hyperlinear approximations. We focus on approximations of Gromov hyperbolic groups, comment open problems, and suggest a conjecture in this setting. Then we turn over this classical way and initiate the study of approximations by groups usually known as being not so elementary to investigate. This allows to see that many interesting groups (still unknown to have algebraic or metric approximations) admit this new type of approximations which we call *asymptotic approximations*. We give many examples of asymptotically sofic/hyperlinear groups, as well as of asymptotically non-residually finite groups. In particular, we provide the first examples of infinite simple asymptotically residually finite (resp. asymptotically amenable) groups with Kazhdan's property (T).

The present text is a transcript of a talk the author gave since 2008 on several occasions, namely at the universities of Neuchâtel, Copenhagen, Aix-Marseille, at the ENS Lyon, ETH Zürich, MF Oberwolfach, and CRM Barcelona. The author is grateful to these institutions for their support and hospitality.

G. Arzhantseva (✉)
Faculty of Mathematics, Universität Wien, Wien, Austria
e-mail: goulnara.arzhantseva@univie.ac.at

J. González-Meneses et al. (eds.), *Extended Abstracts Fall 2012*,
Trends in Mathematics 1, DOI 10.1007/978-3-319-05488-9_2,
© Springer International Publishing Switzerland 2014

2 Classical Idea: Approximate "Complicated" Groups by "Easy" Ones

Let G be a group and S a finite set of generators of G. We denote by $|\cdot|_S$ the word length on G induced by S and by $B_S(n) = \{g \in G : |g|_S \leqslant n\}$ the ball of radius n centered at the identity of G. Let $\mathcal{F} = \{\text{some groups}\}$ be a given family of groups.

Definition 1. A group G is *approximated by* $\mathcal{F}$ if for each $n \in \mathbb{N}$ there exists a map $i \colon B_S(n) \to F \in \mathcal{F}$ such that

$(*)$ $i(g)i(h) = i(gh)$ for all $g, h, gh \in B_S(n)$;
$(**)$ $i(g) \neq i(h)$ for all elements $g \neq h$ of $B_S(n)$.

In other words, G is approximated by $\mathcal{F}$ if the algebraic structures of G and of a group $F \in \mathcal{F}$ coincide whenever we focus on a ball of a given radius in G and its image in F. The map i does depend on n in general but we omit the indexing. We call such a map i an *algebraic approximation*. Assumption $(*)$ is termed to be a *homomorphism on the ball* and $(**)$ an *injectivity on the ball*.

By varying the groups constituting family $\mathcal{F}$ and the choices of the map i, we recognize many famous intensively studied classes of groups. Here are some basic examples.

Examples 2 (Algebraic approximations).

- Residually finite groups (RF) are those approximated by $\mathcal{F} = \{\text{finite groups}\}$ with i a homomorphism $G \to F \in \mathcal{F}$;
- Locally embeddable into finite ones (LEF) are groups approximated by the family $\mathcal{F} = \{\text{finite groups}\}$ [25];
- Residually amenable groups (RA) are groups approximated by the family $\mathcal{F} = \{\text{amenable groups}\}$ with i a homomorphism $G \to F \in \mathcal{F}$;
- Initially subamenable groups (ISA) are groups approximated by the family $\mathcal{F} = \{\text{amenable groups}\}$ [13];
- If G is a fully residually-$\mathcal{F}$ group (FR$\mathcal{F}$), then G is approximated by $\mathcal{F}$;
- If G is a limit of groups from $\mathcal{F}$ ($\lim \mathcal{F}$) in the space of marked groups [11], then G is approximated by $\mathcal{F}$.
- If G is a limit of groups approximated by $\mathcal{F}$ ($\lim A\mathcal{F}$) in the space of marked groups, then G is approximated by $\mathcal{F}$.

The relationship between the above classes and concrete (non)examples of groups are known. For instance, (RF) $\subsetneq$ (LEF) $\subsetneq$ (ISA), (RF) $\subsetneq$ (RA) $\subsetneq$ (ISA) and (FR$\mathcal{F}$) $\subseteq$ ($\lim \mathcal{F}$) $\subseteq$ ($\lim A\mathcal{F}$). The free group $\mathbb{F} = \mathbb{F}(S)$ belongs to all classes, whenever it is residually-$\mathcal{F}$, for the last three examples. Thompson's groups T and V are not (ISA), hence not (RA), etc.

The origin of algebraic approximations goes back to fundamental papers of Malcev and those of Marshall Hall (who seem to come independently to the general concept of residual finiteness; their results have predecessors in works of Schreier and Levi—see [10] and references therein). The development of the topic in group theory has been fruitfully interwoven with that of algebraic topology and spectral geometry.

Nowadays, a major open problem in geometric group theory is the following.

Question 3. Are all Gromov hyperbolic groups residually finite?

We will be back to this question below. Let us now consider more general approximations.

Let $\mathcal{F}_{\text{dist}} = \{\text{some groups with distance}\}$ be a given family of groups, each of which equipped with a bi-invariant distance (assumed to be normalized, for simplicity).

Definition 4. A group G is *metrically approximated by* $\mathcal{F}_{\text{dist}}$ if for each $n \in \mathbb{N}$ there exists a map $\pi : B_S(n) \to (F, \text{dist}) \in \mathcal{F}_{\text{dist}}$ such that

(∗) $\text{dist}(\pi(g)\pi(h), \pi(gh)) < 1/n$ for all $g, h, gh \in B_S(n)$;
(∗∗) $\text{dist}(\pi(g), \pi(h)) > 1 - 1/n$ for all elements $g \neq h$ of $B_S(n)$.

In other words, G is metrically approximated by $\mathcal{F}_{\text{dist}}$ if the algebraic structures of G and of a group $(F, \text{dist}) \in \mathcal{F}_{\text{dist}}$ almost coincide (that is, they differ by a dist-small quantity) whenever we focus on a ball of a given radius in G and its image in (F, dist). The map π does depend on n in general but we omit the indexing. We call such a map π a *metric approximation*. Assumption (∗) is termed to be an *almost homomorphism on the ball* and (∗∗) a *uniform injectivity*.

By varying the groups constituting the family $\mathcal{F}_{\text{dist}}$ and the choices of bi-invariant metrics, we obtain many interesting recently emerged classes of groups.

Examples 5 (Metric approximations).

- Sofic groups (S) are groups metrically approximated by the family $\mathcal{F}_{\text{dist}} = \{\text{Sym}(n), d_{\text{Ham}} \mid n \in \mathbb{N}\}$, where each symmetric group of finite degree is equipped with the normalized Hamming distance d_{Ham}.
- Linear sofic groups (LS) are groups metrically approximated by the family $\mathcal{F}_{\text{dist}} = \{GL_n(\mathbb{C}), d_{\text{rank}} \mid n \in \mathbb{N}\}$, where general linear groups are equipped with the normalized rank distance [4].
- Weakly sofic groups (WS) are groups metrically approximated by the family $\mathcal{F}_{\text{dist}} = \{\text{finite groups with distance}\}$, where each finite group is endowed with a normalized bi-invariant metric [12].
- Hyperlinear groups (H) are groups metrically approximated by the family $\mathcal{F}_{\text{dist}} = \{U(n), d_{\text{HS}} \mid n \in \mathbb{N}\}$, where each unitary group of finite rank is equipped with the normalized Hilbert–Schmidt distance.
- Every group algebraically approximated by $\mathcal{F}_{\text{dist}}$ is metrically approximated by $\mathcal{F}_{\text{dist}}$, provided assumption (∗∗) of Definition 4 holds for the given distance.
- If G is a limit of groups metrically approximated by $\mathcal{F}_{\text{dist}}$ (lim $\mathcal{F}_{\text{dist}}$) in the space of marked groups, then G is metrically approximated by $\mathcal{F}_{\text{dist}}$.

There is much less known on the relationship between the above classes of metrically approximated groups and on the possibility of concrete non-examples. For instance, each of the classes (lim RF), (lim RA), (lim LEF), (lim ISA) is clearly contained in (S) $\subseteq$ (WS). Also, (S) $\subseteq$ (H) as $d_{\mathrm{Ham}}(\sigma, \tau) = d_{\mathrm{HS}}(A_\sigma, A_\tau)^2/2$ for all $\sigma, \tau \in \mathrm{Sym}(n)$ and the corresponding permutation matrices $A_\sigma, A_\tau \in U(n)$. Recently, we have proved that (S) $\subseteq$ (LS) $\subseteq$ (WS); see [4].

The study of metric approximations is motivated by open problems in dynamics and operator algebra. Hyperlinear groups appeared in the context of Alain Connes' embedding conjecture (1976) in operator algebra and were introduced by Florin Rădulescu. Sofic groups were introduced by Misha Gromov in his study of symbolic algebraic varieties in relation to Gottschalk's surjunctivity conjecture (1973) in topological dynamics. They were called sofic by Benjamin Weiss. They are known to satisfy Kaplansky's direct finiteness conjecture (1969) by a result of Elek and Szabo. We refer the reader to nice surveys [22, 23] for more information on sofic and hyperlinear groups.

A major open problem in the theory of metric approximations is the following.

Question 6. Are all groups sofic/linear sofic/weakly sofic/hyperlinear?

We expect a negative answer to both Questions 3 and 6. In this context, we have the following curious result.

Properties (5)—(7) below come from the theory of operator algebras, where they are crucial in relation to Alain Connes' embedding conjecture. Here $\mathcal{S}$ denotes the full group of the hyperfinite aperiodic ergodic measure-preserving equivalence relation and $\mathcal{U} = U(R)$ the unitary group of the hyperfinite factor R of type II_1 equipped with the ultraweak topology. It is known that a group G is sofic (resp. hyperlinear) if and only if it embeds into a metric ultrapower of $\mathcal{S}$ (resp. of $\mathcal{U}$); see [15, 23] for precise definitions.

Proposition 7. *The following are equivalent.*

(1) *All hyperbolic groups are (RF).*
(2) *All hyperbolic groups are (LEF).*
(3) *All hyperbolic groups are (RA).*
(4) *All hyperbolic groups are (ISA).*
(5) *All hyperbolic groups have Kirchberg's factorization property (KFP).*
(6) *All hyperbolic groups can be embedded into $\mathcal{S}$ ($\hookrightarrow \mathcal{S}$).*
(7) *All hyperbolic groups can be embedded into $\mathcal{U}$ ($\hookrightarrow \mathcal{U}$).*

Proof. Since hyperbolic groups are finitely presented, the equivalences (1) $\Longleftrightarrow$ (2) and (3) $\Longleftrightarrow$ (4) are immediate.

Let us show (1) $\Longleftrightarrow$ (3). Finite groups are amenable. Therefore, it suffices to check that (1) $\Longleftarrow$ (3). Assume that all hyperbolic groups are (RA) but there exists a hyperbolic group G_0 which is not (RF). Then, by a result of Ol'shanskii [20] and, independently, of Kapovich and Wise [14], there exists a non-elementary hyperbolic group G with no proper subgroups of finite index. On the other hand, there exists a hyperbolic group G_T which has Kazhdan's property (T). One can take, for instance, a co-compact lattice in $Sp(n, 1), n \geqslant 2$. Such a group is clearly non-elementary. By another result of Ol'shanskii [19], G and G_T has a common quotient Q which is a non-elementary hyperbolic group. Such a group Q has Kazhdan's property (T) by construction and it is (RA) by assumption. Hence, every amenable quotient of Q is finite. This yields a contradiction as Q is infinite and Q has no proper finite index subgroups, by construction.

Implications (1) $\implies$ (5)—(7), as well as (6) $\implies$ (7) and (5) $\implies$ (7) are immediate from the definitions [15, 23]. Let us check that (7) $\implies$ (1). Assume that all hyperbolic groups can be embedded into $\mathcal{U}$ but there exists a hyperbolic group G_0 which is not (RF). Proceeding as above, we find a non-elementary hyperbolic group Q which has Kazhdan's property (T), has no proper subgroups of finite index, and which can be embedded into $\mathcal{U}$. By a result of Kirchberg [15], for a group with Kazhdan's property (T), we have (RF) $\iff$ (KFP) $\iff$ ($\hookrightarrow \mathcal{S}$) $\iff$ ($\hookrightarrow \mathcal{U}$). Thus, Q is (RF). This is a contradiction as Q is infinite and has no proper finite index subgroups, by construction.   □

The preceding proposition can be viewed as a first step to the following equivalence.

Conjecture 8. All hyperbolic groups are residually finite if and only if all hyperbolic groups are sofic.

It is commonly believed that a non-residually finite hyperbolic group does exist. If established, this equivalence indicates a difficulty to find such a counterexample which would answer Questions 3 and 6 in the negative.

3 New Idea: Approximate "Easy" Groups by "Complicated" Ones

The area of metric approximations of discrete groups is very attractive as many natural questions on metrically approximated groups and their applications are open. Various well-studied groups and classes of groups are still not known "*to be, or not to be*" metrically approximated. In particular, the following is unknown.

- Are the following groups sofic/linear sofic/weakly sofic/hyperlinear?

  - Hyperbolic groups and their subgroups.
  - Weakly amenable groups (that is, not uniformly non-amenable groups [5]).
  - One-relator groups.
  - Mapping class groups and outer automorphism groups $\text{Out}(\mathbb{F}_n)$ if $n \geqslant 3$.
  - Thompson's groups F, T, V.

- Does there exist an infinite simple sofic group with Kazhdan's property (T)?

We approach these classes of groups by introducing a new way to approximate finitely generated groups.

Definition 9. A group G is *asymptotically approximated* by $\mathcal{F}_{\text{dist}}$ if for each $n \in \mathbb{N}$ there exist a finite generating set S_n of G and a map $\pi\colon B_{S_n}(n) \to (F, \text{dist}) \in \mathcal{F}_{\text{dist}}$ such that

$(*)$ $\text{dist}(\pi(g)\pi(h), \pi(gh)) < 1/n$ for all $g, h, gh \in B_{S_n}(n)$;

$(**)$ $\text{dist}(\pi(g), \pi(h)) > 1 - 1/n$ for all elements $g \neq h$ of $B_{S_n}(n)$.

In other words, G is asymptotically approximated by $\mathcal{F}_{\text{dist}}$ if there exists a sequence $(S_n)_{n \in \mathbb{N}}$ of finite generating sets of G such that the algebraic structures of G and of a group $(F, \text{dist}) \in \mathcal{F}_{\text{dist}}$ almost coincide (that is, they differ by a *dist-small* quantity) whenever we focus on a ball with respect to S_n of a given radius in G and its image in (F, dist). We call such a map π an *asymptotic approximation*.

Note a reverse order of asymptotic approximation in comparison with a general idea of approximation: a group G is asymptotically approximated by a family of groups $\mathcal{F}_{\text{dist}}$ if the family $\{(G, S_n)\}_{n \in \mathbb{N}}$ approaches, in the above sense, groups from $\mathcal{F}_{\text{dist}}$.

By varying the groups constituting family $\mathcal{F}_{\text{dist}}$ and the choices of bi-invariant metrics, we get the concepts of *asymptotic residual finiteness/residual amenability/soficity/hyperlinearity, etc.* For example, a group asymptotically approximated by a family $\mathcal{F}_{\text{dist}}$ consisting of residually finite groups is called asymptotically residually finite (with respect to the given metrics). In our discussion below, we omit mentioning metrics explicitly as the choice is often rather obvious (e.g., a basic choice: the length of every non-trivial group element is assigned to be 1; relative to the induced metric, assumption $(*)$ means that π is a homomorphism on $B_{S_n}(n)$ and $(**)$ obviously holds).

Examples 10 (Asymptotic approximations).

- All groups metrically approximated by a family $\mathcal{F}_{\text{dist}}$ are asymptotically approximated by $\mathcal{F}_{\text{dist}}$. For instance, residually finite groups are asymptotically residually finite, sofic groups are asymptotically sofic, both classes are asymptotically finite, etc.

- All weakly amenable groups are asymptotically amenable (hence, asymptotically sofic). Indeed, if G is weakly amenable [5], then there exists a sequence $(S_n)_{n \in \mathbb{N}}$ of finite generating sets of G such that $\{(G, S_n)\}_{n \in \mathbb{N}}$ approaches an amenable group in the sense of Definition 9.

  There are numerous examples of non-amenable weakly amenable groups, and many of them are not known to be sofic [5]. Such groups can be made to have rather unusual extreme properties. For instance, there exists a non-amenable weakly amenable simple periodic (hence, with no non-abelian free subgroups) group Q with Kazhdan's property (T). Alternatively, given an arbitrary countable group C, there exists a non-amenable weakly amenable simple group Q with Kazhdan's property (T) such that Q contains an isomorphic copy of C. Other properties can be added to Q (which is a common quotient of all non-elementary hyperbolic groups); see more details on the construction of such a group Q in [6, Propositions 2.6 and 2.8, Remark 2.9] and [5, 21] on their weak amenability.

- Free Burnside groups $B(m, n)$ with $m \geqslant 2$ and odd exponent $n \geqslant 1,003$ are not asymptotically residually finite. Indeed, by the celebrated Novikov–Adyan solution of the bounded Burnside Problem, such a group is infinite and by the famous Zelmanov solution of the restricted Burnside Problem it cannot be residually finite. By a result of Atabekyan [7], there is a number $L < (400\,n)^3$ such that for an arbitrary set K generating a noncyclic subgroup $\langle K \rangle$ of $B(m, n)$, there are elements $u, v \in \langle K \rangle$ for which the pair $\{u, v\}$ is a basis of a free Burnside subgroup of exponent n (which is not residually finite), and the lengths of the elements u and v with respect to the generating set K satisfy the inequalities $|u|_K < L$ and $|v|_K < L$. Thus, if a family $\mathcal{F}_{\mathrm{dist}}$ asymptotically approximates such a $B(m, n)$, then $\mathcal{F}_{\mathrm{dist}}$ cannot consist of residually finite groups only.

The next result provides numerous examples of asymptotically sofic groups (and answers, in this new context of asymptotic approximations, questions from the beginning of this section).

Recall that a girth of a graph is the length of shortest non-trivial loop and the *girth of a group* is the supremum of the girths of all of its Cayley graphs with respect to finite sets of generators [1, 24].

Theorem 11. *All groups of infinite girth are asymptotically residually finite. In particular, the following groups are asymptotically residually finite (hence, asymptotically sofic):*

- *Hyperbolic groups and their finitely generated subgroups.*
- *One-relator groups.*
- *Thompson's group F; more generally, finitely generated subgroups of $PL_o(I)$.*
- *Finitely generated subgroups of convergence groups.*
- *Finitely generated subgroups of a mapping class group.*
- *Outer automorphism groups $\mathrm{Out}(\mathbb{F}_n)$ and its subgroups with an (iwip) element.*

In addition, there exists an infinite simple asymptotically residually finite (hence, asymptotically sofic) group with Kazhdan's property (T).

Proof. By definition, groups of infinite girth are asymptotically approximated by a family consisting of a free non-abelian group $\mathbb{F}$ of finite rank. Therefore, they are asymptotically residually finite (or adapting our terminology, *asymptotically free*). The groups above are indeed asymptotically residually finite as they are known to be either of infinite girth or residually finite.

Hyperbolic groups and their finitely generated subgroups are of infinite girth [1] whenever they are not virtually cyclic (otherwise, they are obviously residually finite).

One-relator groups are of infinite girth [1] if and only if they are not solvable. Virtually solvable subgroups of one-relator groups consist of: cyclic groups of finite or infinite order, free abelian groups of rank 2, the fundamental group of a Klein bottle, and the Baumslag–Solitar group $BS(1, n) = \langle a, b \mid aba^{-1} = b^n \rangle$ with $|n| \geqslant 2$. All these groups are residually finite.

Thompson's group F is of infinite girth [3, 9], therefore it is asymptotically free (hence, asymptotically residually finite). For finitely generated subgroups of $PL_o(I)$, the group of orientation preserving piecewise linear homeomorphisms of the closed interval, see [2], where non-solvable subgroups are proven to be of infinite girth. Solvable subgroups of $PL_o(I)$ are characterized in [8]—they are all residually finite.

The results on the convergence (resp. the mapping class) groups follow from [17, 26], as such subgroups are of infinite girth whenever they are not virtually cyclic (resp. virtually abelian), and similarly for subgroups of $Out(\mathbb{F}_n)$ [16] having a so-called (iwip) element. The group $Out(\mathbb{F}_n)$ itself has infinite girth in an obvious way as it surjects onto $GL_n(\mathbb{Z})$, which is of infinite girth [1].

By a result of Ol'shanskii, there exists a torsion-free Tarski monster group, that is, an infinite non-abelian group all whose proper subgroups are infinite cyclic [18, Ch. 9, §28.1]. Moreover, there exists such a group that does not satisfy any non-trivial identity, hence of infinite girth; see Ol'shanskii's argument and the proof that such a group is of infinite girth in [27]. Clearly, Tarski's monster is simple. In addition, it can be made to have Kazhdan's property (T): it suffices to build such a torsion-free Tarski monster starting from a torsion-free hyperbolic group with property (T)—for instance, a co-compact lattice in $Sp(n, 1), n \geqslant 2$ [19].   □

We do not know any example of a group which is not asymptotically sofic/hyper-linear (e.g., for a basic choice of the metrics).

Acknowledgements This research was supported in part by ERC grant ANALYTIC no. 259527 and by the Swiss NSF under Sinergia grant CRSI22-130435.

References

1. A. Akhmedov, On the girth of finitely generated groups. J. Algebra **268**(1), 198–208 (2003)
2. A. Akhmedov, Girth alternative for subgroups of $PL_o(I)$ (2011). arXiv:1105.4908
3. A. Akhmedov, M. Stein, J. Taback, Free limits of Thompson's group F. Geom. Dedic. **155**, 163–176 (2011)
4. G. Arzhantseva, L. Păunescu, Linear sofic groups and algebras (2012). arXiv:1212.6780
5. G. Arzhantseva, J. Burillo, M. Lustig, L. Reeves, H. Short, E. Ventura, Uniform non-amenability. Adv. Math. **197**(2), 499–522 (2005)
6. G. Arzhantseva, M. Bridson, T. Januszkiewicz, I. Leary, A. Minasyan, J. Świątkowski, Infinite groups with fixed point properties. Geom. Topol. **13**(3), 1229–1263 (2009)
7. V. Atabekyan, Uniform nonamenability of subgroups of free Burnside groups of odd period. Mat. Zametki **85**(4), 516–523 (2009); English translation: Math. Notes **85**(3–4), 496–502 (2009)
8. C. Bleak, An algebraic classification of some solvable groups of homeomorphisms. J. Algebra **319**(4), 1368–1397 (2008)
9. M. Brin, The free group of rank 2 is a limit of Thompson's group F. Groups Geom. Dyn. **4**(3), 433–454 (2010)
10. R. Campbell, Residually finite groups, PhD thesis, 1989, unpublished. http://userpages.umbc. edu/~rcampbel/CombGpThy/RF_Thesis/index.html

11. C. Champetier, V. Guirardel, Limit groups as limits of free groups. Isr. J. Math. **146**, 1–75 (2005)
12. L. Glebsky, L.M. Rivera, Sofic groups and profinite topology on free groups. J. Algebra **320**(9), 3512–3518 (2008)
13. M. Gromov, Endomorphisms of symbolic algebraic varieties. J. Eur. Math. Soc. **1**(2), 109–197 (1999)
14. I. Kapovich, D. Wise, The equivalence of some residual properties of word-hyperbolic groups. J. Algebra **223**(2), 562–583 (2000)
15. E. Kirchberg, Discrete groups with Kazhdan's property T and factorization property are residually finite. Math. Ann. **299**(3), 551–563 (1994)
16. K. Nakamura, Some results in topology and group theory. PhD thesis, University of California, Davis, 2008
17. K. Nakamura, The girth alternative for mapping class groups (2011). arXiv:1105.5422
18. A.Yu. Ol'shanskiĭ, *Geometry of Defining Relations in Groups* (Kluwer, Dordrecht/Boston, 1991)
19. A.Yu. Ol'shanskiĭ, On residualing homomorphisms and G-subgroups of hyperbolic groups. Int. J. Algebra Comput. **3**(4), 365–409 (1993)
20. A.Yu. Ol'shanskiĭ, On the Bass-Lubotzky question about quotients of hyperbolic groups. J. Algebra **226**(2), 807–817 (2000)
21. D. Osin, Weakly amenable groups, in *Computational and Statistical Group Theory*, Las Vegas/Hoboken, 2001. Contemporary Mathematics, vol. 298 (American Mathematical Society, Providence, 2002), pp. 105–113
22. V. Pestov, Hyperlinear and sofic groups: a brief guide. Bull. Symb. Log. **14**(4), 449–480 (2008)
23. V. Pestov, A. Kwiatkowska, An introduction to hyperlinear and sofic groups (2009). arXiv: 0911.4266
24. S. Schleimer, On the girth of groups (2000, preprint). Available at http://homepages.warwick.ac.uk/~masgar/Maths/girth.pdf
25. A. Vershik, E. Gordon, Groups that are locally embeddable in the class of finite groups. Algebra i Analiz **9**(1), 71–97 (1997); English translation: St. Petersb. Math. J. **9**(1), 49–67 (1998)
26. S. Yamagata, The girth of convergence groups and mapping class groups. Osaka J. Math. **48**(1), 233–249 (2011)
27. P. Zusmanovich, On the utility of Robinson-Amitsur ultrafilters (2009). arXiv:0911.5414

An Efficient Algorithm for Finding a Basis of the Fixed Point Subgroup of an Automorphism of a Free Group

Oleg Bogopolski and Olga Maslakova

Let F_n be the free group of finite rank n. For $\alpha \in \mathrm{Aut}(F_n)$ we set

$$\mathrm{Fix}(\alpha) = \{x \in F_n \mid \alpha(x) = x\}.$$

In [1], Bestvina and Handel proved the Scott conjecture that $\mathrm{rk}\,\mathrm{Fix}(\alpha) \leqslant n$. However, the problem of finding a basis of $\mathrm{Fix}(\alpha)$ has been open for almost 20 years. It has been solved in three special cases: for positive automorphisms [5], for special irreducible automorphisms [9, Proposition B], and for all automorphisms of F_2 [2]. In 1999, Maslakova, a former PhD student of the first-named author, attempted to solve this problem in full generality. However, her proof published in [7] was not complete. An improved, but still incomplete proof was given in her PhD thesis [8]. So, we have decided to give a full and correct proof. The main result of our paper [3] is:

Theorem 1. *Let F_n be the free group of finite rank n. There is an efficient algorithm which, given an automorphism α of F_n, finds a basis of the fixed point subgroup $Fix(\alpha)$.*

Here we give a sketch of the proof. We assume that the reader is familiar with the relative train track technique from [1].

O. Bogopolski (✉)
Mathematisches Institut, Universität Düsseldorf, Düsseldorf, Germany
e-mail: bogopolski@math.uni-duesseldorf.de

O. Maslakova
Sobolev Institute of Mathematics, Siberian Branch of the Russian Academy of Sciences, Novosibirsk, Russia
e-mail: o.s.maslakova@gmail.com

J. González-Meneses et al. (eds.), *Extended Abstracts Fall 2012*,
Trends in Mathematics 1, DOI 10.1007/978-3-319-05488-9_3,
© Springer International Publishing Switzerland 2014

1   A Modification of a Relative Train Track for α

Let Γ be a finite graph and p be a path in Γ. The initial point of p is denoted by $\iota(p)$ and the terminal one by $t(p)$. The homotopy class of p relative to the endpoints of p is denoted by $[[p]]$, and the unique reduced path from $[[p]]$ is denoted by $[p]$. The trivial path at a point v is denoted by $\mathbf{1}_v$.

Let $f : \Gamma \to \Gamma$ be a relative train track and $\varnothing = \Gamma_0 \subset \cdots \subset \Gamma_N = \Gamma$ be the maximal filtration for f. Each stratum $H_i := \mathrm{cl}(\Gamma_i \setminus \Gamma_{i-1})$ is either exponential, polynomial, or zero; for each exponential stratum H_r there is a real number $\lambda_r > 1$ and a pseudo-metric L_r on Γ_r such that

- If E is an edge in Γ_r, then $L_r(E) > 0$ if and only if E is an edge in H_r;
- If p is a reduced r-legal path in Γ_r, then $L_r([f(p)]) = \lambda_r L_r(p)$.

Observe that for computing a basis of $\mathrm{Fix}(\alpha)$, it suffices to compute a basis of $\mathrm{Fix}(\alpha^m)$ for some $m \geqslant 1$. So we can replace α by an appropriate power of α if needed.

Theorem 2. *Replacing α by an appropriate α^m, we can construct a relative train track $f : (\Gamma, v) \to (\Gamma, v)$ and indicate an isomorphism $i : F \to \pi_1(\Gamma, v)$ such that $i^{-1}\alpha i = f_*$ (in this case we say that f topologically represents α) and the following condition is satisfied:*

(**Pol**) *Every polynomial stratum H_r consists of only two mutually inverse edges, say E and $\overline{E}$. Moreover, $f(E) \equiv E \cdot a$, where a is a path in Γ_{r-1}.*

We say that an edge-path p in Γ is of *height* r if it lies in Γ_r and has at least one edge in H_r.

Definition 3. Let μ be a reduced edge-path of height r in Γ, where H_r is an exponential stratum. Let y be an occurrence of a vertex in μ. The vertex y is called an *r-cancellation point* in μ if the turn at y in μ is an illegal r-turn.

Let $y_1, \ldots, y_k$ be all r-cancellation points in μ. We say that these points are *non-deletable in μ* if the number of r-cancellation points in $[f^i(\mu)]$ is equal to k for each $i \geqslant 0$. In this case the path μ is called *r-stable*.

Suppose that μ is r-stable and $y_1, \ldots, y_k$ are all r-cancellation points in μ. Let $y_0 = \iota(\mu)$ and $y_{k+1} = t(\mu)$, and let μ_i be the subpath of μ between y_i and y_{i+1}; thus $\mu \equiv \mu_0 \mu_1 \ldots \mu_k$.

Let $i \in \{1, \ldots, k\}$. The *r-cancellation area around y_i in μ*, denoted $A(\mu, y_i)$, is the maximal subpath of $\mu_{i-1}\mu_i$ which contains y_i and consists of points u of $\mu_{i-1}\mu_i$ which cancel out by reducing the product $[f^\ell(\mu_{i-1})] \cdot [f^\ell(\mu_i)]$ for appropriate $\ell = \ell(u) \geqslant 0$. We stress that $A(\mu, y_i)$ is not necessarily an edge-path.

If μ is r-stable and $y_1, \ldots, y_k$ are all non-deletable r-cancellation points in μ, then we can write μ in the form $\mu \equiv b_0 A_1 b_1 \ldots A_k b_k$, where the paths b_j are r-legal or trivial and $A_1, \ldots, A_k$ are r-cancellation areas around $y_1, \ldots, y_k$ in μ. This form is called the *A-decomposition* of μ.

A path p in Γ is called an *r-cancellation area* in Γ if it has the form $A(\mu, y)$ for some r-stable path μ and a non-deletable r-cancellation point y in μ.

Lemma 4. 1. *There exist only finitely many r-cancellation areas in Γ and they can be computed.*
2. *Given μ as in Definition 3, one can compute $s \geq 0$ such that $[f^s(\mu)]$ is r-stable. In particular, one can verify whether the r-cancellation points in μ are non-deletable or not.*
3. *If μ is r-stable and has the A-decomposition $\mu \equiv b_0 A_1 b_1 \cdots A_k b_k$, then, for every $i \geq 1$, $[f^i(\mu)]$ has the A-decomposition*

$$[f^i(\mu)] \equiv [f^i(b_0)][f^i(A_1)][f^i(b_1)] \cdots [f^i(A_k)][f^i(b_k)].$$

Theorem 5. *Let $f : (\Gamma, v) \to (\Gamma, v)$ be the relative train track from Theorem 2. Using subdivisions, one can obtain a new relative train track $f' : (\Gamma', v) \to (\Gamma', v)$ topologically representing α and satisfying (Pol) and*

(RTT-iv) *There is a computable natural number $P = P(f')$ such that, for each exponential stratum H_r and each r-cancellation area A in Γ', the r-cancellation area $[(f')^P(A)]$ is an edge-path.*

The map f' is called the *subdivided relative train track*.

2 An Auxiliary Graph D_f

Let Γ be a finite connected graph with a distinguished vertex v and let $f : \Gamma \to \Gamma$ be a homotopy equivalence such that $f(\Gamma^0) \subseteq \Gamma^0$, f is locally injective on $\Gamma \setminus \Gamma^0$, and $f(v) = v$. We define a graph D_f (first introduced in [6]) and describe a procedure (see [9] and [5] for special cases) which helps to compute a basis of the group

$$\overline{\mathrm{Fix}}(f) := \{[[p]] \in \pi_1(\Gamma, v) \mid f(p) = p\}.$$

In Sect. 3 we sketch how to convert this procedure into an algorithm in the case where f is a subdivided relative train track. Due to Theorem 5, $\mathrm{Fix}(\alpha)$ is naturally isomorphic to $\overline{\mathrm{Fix}}(f)$, so this will give us an algorithm for computing a basis of $\mathrm{Fix}(\alpha)$.

Definition 6 (of the graph D_f). An *f-path* in Γ is an edge-path μ in Γ (possibly trivial) such that f maps $\iota(\mu)$ to $t(\mu)$. Thus, if μ is an f-path in Γ, the path $\mu f(\mu)$ is well defined. The vertices of the graph D_f are reduced f-paths. A vertex μ of D_f is called *trivial* if $\mu = \mathbf{1}_u$ for some vertex u of Γ fixed by f. Two vertices μ and τ in D_f are connected by an edge (from μ to τ) with label E if E is an edge in Γ outgoing from the same vertex as μ and $[\overline{E}\mu f(E)] = \tau$.

We set $\hat{f}(\mu) := [\overline{E}\mu f(E)]$ if E is the first edge of the f-path μ. Clearly, μ and $\hat{f}(\mu)$ are connected in D_f by an edge with the label E. The direction of this edge is called *preferable* at μ. Preferable directions at all nontrivial vertices determine a flow in D_f. Starting at μ and moving along this flow, we get vertices $\mu = \mu_1, \mu_2, \ldots$, where $\mu_{i+1} = \hat{f}(\mu_i)$, $i \geq 1$. These vertices together with the directed edges we pass form a subgraph in D_f which we call the μ-*subgraph*. The μ-subgraph is either a finite segment, or a finite segment with a cycle, or a ray.

An edge e connecting two vertices u, w in D_f is called *repelling* if the direction of e is not preferable at u and the direction of the opposite edge $\overline{e}$ is not preferable at w. Endpoints of repelling edges are called *repelling vertices*. There exist only finitely many repelling edges in D_f and they can be algorithmically found; see [6, 9] or [5, Lemma 3.5].

Recall that v is the distinguished vertex of Γ and $f(v) = v$. Since $\mathbf{1}_v$ is an f-path, we can consider $\mathbf{1}_v$ as a vertex of D_f. Let $D_f(\mathbf{1}_v)$ be the component of D_f containing $\mathbf{1}_v$.

Lemma 7 (see [6]). *The fundamental group $\pi_1(D_f(\mathbf{1}_v), \mathbf{1}_v)$ can be identified with $\overline{Fix}(f)$ through the map $[[p]] \mapsto [[l(p)]]$, where p is a closed path in $D_f(\mathbf{1}_v)$ based at $\mathbf{1}_v$ and $l(p)$ is its label.*

A component of D_f is called *repelling* if it contains at least one repelling edge. Let $C_1, \ldots, C_m$ be all repelling components of D_f. For each C_i, let $CoRe(C_i)$ be a core of C_i which contains all repelling edges of C_i. We set $C_f := \bigcup_{i=1}^m C_i$ and $CoRe(C_f) := \bigcup_{i=1}^m CoRe(C_i)$.

Lemma 8. 1. *The vertex $\mathbf{1}_v$ lies in C_f iff it lies in the μ-subgraph for some repelling vertex μ of C_f.*
2. *If $\mathbf{1}_v$ does not lie in C_f, then $D_f(\mathbf{1}_v)$ is contractible.*
3. *Suppose that $\mathbf{1}_v$ lies in C_f and let μ be the repelling vertex from (1). Let Δ be the union of the μ-subgraph and the component of $CoRe(C_f)$ containing μ. Then the inclusion of Δ in $D_f(\mathbf{1}_v)$ induces an isomorphism $\pi_1(\Delta, \mathbf{1}_v) \to \pi_1(D_f(\mathbf{1}_v), \mathbf{1}_v)$.*

Lemma 9. *To compute a basis of $\pi_1(D_f(\mathbf{1}_v), \mathbf{1}_v)$, it suffices to construct $CoRe(C_f)$ and to decide whether the vertex $\mathbf{1}_v$ lies in the μ-subgraph for some repelling vertex μ.*

It turns out that $CoRe(C_f)$ is contained in the union of the repelling edges and the μ-subgraphs, where μ runs over the set of repelling vertices. So, to construct $CoRe(C_f)$, it suffices to do the following:

1. Compute repelling edges.
2. For each repelling vertex μ, determine whether the μ-subgraph is finite or not.
3. Compute all elements of all finite μ-subgraphs from (2).
4. For each two repelling vertices μ and τ with infinite μ-and τ-subgraphs, determine whether these subgraphs intersect.
5. If the μ-subgraph and the τ-subgraph from (4) intersect, find their first intersection point and compute their initial segments up to this point.

As it was mentioned above, step (1) can be done algorithmically. It turns out that steps (2)–(5) can be done algorithmically if the following two problems are solvable:

Membership problem. Given two vertices μ and τ of the graph D_f, verify whether τ is contained in the μ-subgraph.

Finiteness problem. Given a vertex μ of the graph D_f, determine whether the μ-subgraph is finite or not.

3   Solution of the Membership and Finiteness Problems

From now on we assume that $f: (\Gamma, v) \to (\Gamma, v)$ is a subdivided relative train track; in particular, f satisfies (Pol) and (RTT-iv). Let $\varnothing = \Gamma_0 \subset \cdots \subset \Gamma_N = \Gamma$ be the filtration associated with f.

Definition 10. 1. Let H_r be an exponential stratum. A reduced f-path $\mu \subset \Gamma_r$ is called *r-perfect* if

- The first edge of μ belongs to H_r;
- μ is r-legal;
- $[\mu f(\mu)] \equiv \mu \cdot [f(\mu)]$ and the turn of this path at the point between μ and $[f(\mu)]$ is legal.

2. Let H_r be an exponential stratum. A reduced f-path $\mu \subset \Gamma_r$ containing edges from H_r is called *A-perfect* if

- μ is r-stable and all r-cancellation areas in μ are edge-paths;
- The A-decomposition of μ begins with an A-area, i.e., it has the form $\mu \equiv A_1 b_1 \cdots A_k b_k$;
- $[\mu f(\mu)] \equiv \mu \cdot [f(\mu)]$ and the turn at the point between μ and $[f(\mu)]$ is legal.

3. Let H_r be a polynomial stratum (so, it contains only one, up to inversion, edge E). A reduced f-path $\mu \subset \Gamma_r$ containing edges from H_r is called *E-perfect* if

- The first edge of μ is E or $\overline{E}$;
- Each path $\hat{f}^{\,i}(\mu)$, $i \geqslant 1$, contains the same number of E-edges as μ.

   A vertex v of D_f is called r-perfect or A-perfect, or E-perfect if v considered as an f-path in Γ is r-perfect or A-perfect, or E-perfect, respectively.

Remark 11. If $\mu \equiv E_1 E_2 \cdots E_m$ is an r-perfect or an A-perfect f-path, then there is no cancellation in passing from μ to $\hat{f}(\mu) \equiv E_2 E_3 \cdots E_m \cdot f(E_1)$.

Lemma 12. *Given a vertex μ in D_f, we can efficiently find a vertex v_0 in the μ-subgraph such that one of the following is satisfied:*

(1) *The v_0-subgraph is finite (then the μ-subgraph is finite);*
(2) *v_0 is either r-perfect, or A-perfect, or E-perfect.*

   In case (1), we can efficiently compute all vertices of the μ-subgraph.

In particular, moving along the μ-subgraph, we can detect one of:

- The μ-subgraph is finite;
- The μ-subgraph contains a perfect vertex v_0.

In the latter case, we still have to decide whether the μ-subgraph is finite or not.

Subcase 1. Suppose that v_0 is r-perfect. In this subcase we prove the following:

(1) $L_r(\hat{f}^{i+1}(v_0)) \geqslant L_r(\hat{f}^i(v_0)) > 0$ for all $i \geqslant 0$.
(2) There exist computable natural numbers $m_1 < m_2 < \cdots$ with the property that
    $L_r(\hat{f}^{m_i}(v_0)) = \lambda_r^i L_r(v_0)$ for all $i \geqslant 1$.

This implies that the μ-subgraph is infinite and the membership problem for the μ-subgraph is solvable.

Subcase 2. Suppose that v_0 is A-perfect. Then we prove that it is possible to find a finite set $\{v_0, v_1, \ldots, v_k\}$ of A-perfect vertices in the v_0-subgraph such that all A-perfect vertices in the v_0-subgraph are:

$$v_0, \quad v_1, \quad \ldots, v_k,$$
$$[f(v_0)], \quad [f(v_1)], \quad \ldots, [f(v_k)],$$
$$[f^2(v_0)], [f^2(v_1)], \ldots, [f^2(v_k)], \ldots$$

Moreover, given a vertex u in the v_0-subgraph, we can find ℓ such that $\hat{f}^\ell(u)$ is an A-perfect vertex.

Using this, the finiteness and the membership problems for the v_0-subgraph can be reduced to:

Problem FIN(v_0). Does there exist $p > q \geqslant 0$ such that $[f^p(v_0)] = [f^q(v_0)]$?

Problem MEM(v_0). Given an f-path τ, does there exist $p \geqslant 0$ with $[f^p(v_0)] = \tau$? Both can be answered with the help of a theorem of Brinkmann; see [4, Theorem 0.1].

Subcase 3. Suppose v_0 is E-perfect. This is treated similarly to Subcase 1; see [3].

References

1. M. Bestvina, M. Handel, Train tracks and automorphisms of free groups. Ann. Math. **135**(1), 1–53 (1992)
2. O. Bogopolski, Classification of automorphisms of the free group of rank 2 by ranks of fixed-point subgroups. J. Group Theory **3**(3), 339–351 (2000)
3. O. Bogopolski, O. Maslakova, A basis of the fixed point subgroup of an automorphism of a free group (2012, preprint). Available at arxiv.org/pdf/1204.6728.pdf
4. P. Brinkmann, Detecting automorphic orbits in free groups. J. Algebra **324**, 1083–1097 (2010)

5. M.M. Cohen, M. Lustig, On the dynamics and the fixed subgroup of a free group automorphism. Invent. Math. **96**(3), 613–638 (1989)
6. R.Z. Goldstein, E.C. Turner, Fixed subgroups of homomorphisms of free groups. Bull. Lond. Math. Soc. **18** 468–470 (1986)
7. O.S. Maslakova, Fixed point subgroup of an automorphism of a free group. Algebra Log. **42**(4), 422–472 (2003) (in Russian)
8. O.S. Maslakova, Fixed point subgroup of an automorphism of a free group. Ph.D. thesis, 2004. www.dslib.net/mat-logika/gruppa-nepodvizhnyh-tochek-avtomorfizma-svobodnoj-gruppy.html
9. E.C. Turner, Finding indivisible Nielsen paths for a train tracks map, in *Combinatorial and Geometric Group Theory: Proceedings of a Workshop*, Held at Heriot-Watt University, Edinburgh, 1993. London Mathematical Society Lecture Note Series, vol. 204 (Cambridge University Press, Cambridge, 1995), pp. 300–313

Metric Properties and Distortion in Nilpotent Groups

José Burillo and Eric López Platón

1 Introduction

One of the concepts developed to study groups from the metric point of view is the concept of distortion of a subgroup in a group. The concept appears already in Gromov's paper [2], and has been studied by several authors, such as Bridson [1], or Sapir and Ol'shanskii (see [3] and [4]). In this paper we study distortion functions obtained by several nilpotent groups embedded into each other. The two families studied are Heisenberg groups and the groups of unipotent upper-triangular matrices. As it could be expected, distortion is polynomial in all cases, and precise degrees are computed for different embeddings between them. The main tool to study this distortion are the estimates of the metric, quantities that can be easily computed for a given element (and its normal form) and which differ from the actual metric by a multiplicative constant. Hence, these estimates are sufficient to compute the distortion functions, and allow us to obtain precise values for them. Some results in this paper coincide with results obtained by D. Osin in [5].

2 Definitions

Definition 1. Let G be a finitely generated group and $H < G$ a subgroup, also finitely generated. Define the *distortion of H in G* as

$$\Delta_H^G(n) = \max \left\{ \|x\|_H \ : \ x \in H, \ \|x\|_G \leq n \right\}.$$

J. Burillo (✉) • E.L. Platón
Departament de Matemàtica Aplicada IV, Universitat Politècnica de Catalunya, Barcelona, Spain
e-mail: burillo@ma4.upc.edu; eric.lopez@ma4.upc.edu

J. González-Meneses et al. (eds.), *Extended Abstracts Fall 2012*,
Trends in Mathematics 1, DOI 10.1007/978-3-319-05488-9_4,
© Springer International Publishing Switzerland 2014

Definition 2. Given a finitely generated group G, an *estimate of the metric* or a *quasi-metric* is a map $E_G \colon G \longrightarrow \mathbb{N}$ such that there exist two constants $C, D > 0$ for which

$$\frac{E_G(x)}{C} - D \le \|x\|_G \le C\, E_G(x) + D$$

for every $x \in G$.

3 Heisenberg Groups

For the $(2k+1)$-dimensional Heisenberg group $\mathcal{H}_{2k+1}$, we have the following quasi-metric:

Theorem 3. *The map* $E \colon \mathcal{H}_{2k+1} \longrightarrow \mathbb{N}$ *defined by*

$$E \begin{pmatrix} 1 & n_1 & n_2 & \cdots & n_k & p \\ & 1 & 0 & \cdots & 0 & m_1 \\ & & 1 & \cdots & 0 & m_2 \\ & & & \ddots & \vdots & \vdots \\ & & & & 1 & m_k \\ & & & & & 1 \end{pmatrix} = \sum_{i=1}^{k} |n_i| + \sum_{j=1}^{k} |m_j| + \sqrt{|p|}$$

is a quasi-metric for $\mathcal{H}_{2k+1}$.

Corollary 4. *All cyclic subgroups of* $\mathcal{H}_{2k+1}$ *are undistorted except those in the center, i.e., generated by a matrix where all* $m_i = n_i = 0$, *which are distorted quadratically.*

It is clear that if $k \le l$, we have embeddings of $\mathcal{H}_{2k+1}$ in $\mathcal{H}_{2l+1}$. For instance, if we take a subset $K \subset \{1, 2, \dots, l\}$ where K has cardinal k, the subgroup of $\mathcal{H}_{2l+1}$ where $n_i = m_i = 0$ for $i \notin K$ is a copy of $\mathcal{H}_{2k+1}$.

Theorem 5. *If* $k \le l$, *the natural embeddings of* $\mathcal{H}_{2k+1}$ *into* $\mathcal{H}_{2l+1}$ *are undistorted.*

4 Upper-Triangular Matrices

Let $\mathcal{T}_n$ be the group of unipotent upper-triangular matrices with integer coefficients. Clearly, $\mathcal{H}_3 = \mathcal{T}_3$. Our goal is to estimate its metric as we have done for $\mathcal{H}_{2k+1}$, and find the distortion function of the different embeddings.

Theorem 6. *The map* $E: \mathcal{T}_n \longrightarrow \mathbb{N}$ *defined by*

$$
E
\begin{pmatrix}
1 & m_{12} & m_{13} & \cdots & m_{1,n-1} & m_{1n} \\
 & 1 & m_{23} & \cdots & m_{2,n-1} & m_{2n} \\
 & & 1 & \cdots & m_{3,n-1} & m_{3n} \\
 & & & \ddots & \vdots & \vdots \\
 & & & & 1 & m_{n-1,n} \\
 & & & & & 1
\end{pmatrix}
= \sum_{1 \leq i < j \leq n} |m_{ij}|^{\frac{1}{j-i}}
$$

is a quasi-metric for $\mathcal{T}_n$.

Again we can deduce from this estimate the distortion of a cyclic subgroup:

Corollary 7. *A cyclic subgroup of* $\mathcal{T}_k$ *generated by a matrix* X *has distortion* n^{j-i}, *where* $j - i$ *is the smallest quantity for which* m_{ij} *is nonzero, i.e.,* m_{ij} *is the nonzero entry which is closest to the diagonal in* X.

The main distortion theorem is:

Theorem 8. *The distortion function for* $\mathcal{H}_{2k+1}$ *as a subgroup of* $\mathcal{T}_{k+2}$ *is polynomial of degree* k.

Finally, we can study the distortion of the different embeddings of $\mathcal{T}_k$ inside $\mathcal{T}_l$ for $k < l$.

Theorem 9. *Given positive integers* $k < l$, *for each* r *such that* $1 \leq r \leq l - k + 1$ *there is an embedding of* $\mathcal{T}_k$ *inside* $\mathcal{T}_l$ *with distortion* n^r.

In particular, observe that inside $\mathcal{T}_{k+1}$ there are subgroups isomorphic to $\mathcal{T}_k$ which are undistorted, and some other subgroups $\mathcal{T}_k$ which are quadratically distorted.

References

1. M.R. Bridson, Fractional isoperimetric inequalities and subgroup distortion. J. Am. Math. Soc. **12**(4), 1103–1118 (1999)
2. M. Gromov, *Asymptotic Invariants of Infinite Groups, Geometry Group Theory, Vol. 2*, Sussex, 1991. London Mathematical Society Lecture Note Series, vol. 182 (Cambridge University Press, Cambridge, 1993), pp. 1–295
3. A.Y. Ol'shanskiĭ, On the distortion of subgroups of finitely presented groups. Math. Sb. **188**(11), 51–98 (1997)
4. A.Y. Ol'shanskiĭ, M.V. Sapir, Length and area functions on groups and quasi-isometric Higman embeddings. Int. J. Algebra Comput. **11**(2), 137–170 (2001)
5. D.V. Osin, Subgroup distortions in nilpotent groups. Comm. Algebra **29**(12), 5439–5463 (2001)

Coset Intersection Graphs, and Transversals as Generating Sets for Finitely Generated Groups

Jack Button, Maurice Chiodo, and Mariano Zeron-Medina Laris

1 Coset Intersection Graphs

Recall that if G is a finitely generated group, then the *rank* of G, $d(G)$, is the minimal size of a generating set for G.

Definition 1. Let G be a group and H a subgroup of G. A *left transversal* for H in G is a set $\{t_\alpha\}_{\alpha \in I} \subseteq G$ such that for each left coset gH there is precisely one $\alpha \in I$ satisfying $t_\alpha H = gH$. A *right transversal* for H in G is defined in an analogous fashion. A *left-right* transversal for H is a set S which is simultaneously a left transversal and a right transversal for H in G.

To understand how left and right cosets behave in a group, we introduce the notion of a coset intersection graph.

Definition 2. Let G be a group and H, K subgroups of G. We define the *coset intersection graph* $\Gamma_{H,K}^G$ to be a graph with vertex set consisting of all left cosets of H ($\{l_i H\}_{i \in I}$) together with all right cosets of K ($\{Kr_j\}_{j \in J}$), where I, J are index sets. If a left coset of H and a right coset of K correspond, they are still included twice. Edges (undirected) are included whenever any two of these cosets intersect.

J. Button (✉)
Selwyn College, University of Cambridge, Cambridge, UK
e-mail: J.O.Button@dpmms.cam.ac.uk

M. Chiodo
Dipartimento di Matematica 'Federigo Enriques', Università degli Studi di Milano, Milano, Italy
e-mail: maurice.chiodo@unimi.it

M.Z.-M. Laris
Department of Pure Mathematics and Mathematical Statistics, Centre for Mathematical Sciences,
University of Cambridge, Cambridge, UK
e-mail: marianozeron@gmail.com

J. González-Meneses et al. (eds.), *Extended Abstracts Fall 2012*,
Trends in Mathematics 1, DOI 10.1007/978-3-319-05488-9__5,
© Springer International Publishing Switzerland 2014

Observing that left (respectively, right) cosets do not intersect, we see that this graph immediately demonstrates some obvious symmetry.

Lemma 3. *Let $H, K < G$. Then $\Gamma^G_{H,K}$ is a bipartite graph, split between $\{l_i H\}_{i \in I}$ and $\{Kr_j\}_{j \in J}$.*

What is less obvious, but still follows from elementary group theory, is the following symmetry of this graph.

Theorem 4. *Let $H, K < G$. Then $\Gamma^G_{H,K}$ is a disjoint union of complete bipartite graphs.*

As soon as we impose some finiteness conditions on the subgroups (finite index or finite size), then this graph exhibits a remarkable level of symmetry.

Theorem 5. *Let $H, K < G$. Suppose that either $|H| = m$, $|K| = n$ (where both subgroups are finite), or $|G : H| = n$, $|G : K| = m$ (where both subgroups have finite index). Then the graph $\Gamma^G_{H,K}$ is a collection of disjoint, complete, finite, bipartite graphs, where each component is of the form K_{s_i, t_i} with $s_i / t_i = n/m$.*

As an immediate corollary, we recover Hall's theorem for transversals:

Corollary 6. *Let $H, K < G$ be of finite index with $|G : H| = |G : K| = n$. Then there exists a set $T \subseteq G$ which is both a left transversal for H in G and a right transversal for K in G. If $H = K$ in G, then T becomes a left-right transversal for H.*

A historical remark, researched by Warren Dicks: The results in this section have a somewhat piecemeal historical origin. A weaker version of Corollary 6, that a subgroup of a finite group always has a left-right transversal, appeared in 1910 by Miller [4, Equation B]. In 1913, Chapman [1] proved the same result; he then realised the existence of the proof by Miller and in 1914 issued a corrigendum [2]. In 1927, Scorza [7] proved Corollary 6 for two separate subgroups H, K, but still assuming G was finite. By the time of Zassenhaus' text [9] in 1937, Corollary 6 was known for finite index subgroups of infinite groups. In 1941, Shü [8] addressed this problem in a way that leaves us somewhat confused. Then in 1958 Ore [5] expanded significantly on such ideas and gave what is to-date the most complete treatment of these. Although our Theorems 4 and 5 can be derived from Ore's work, the symmetry exhibited by the coset intersection graph is not immediately obvious from his terminology. Our methods are completely self-contained, give a very clear picture of what is happening, and require less than one page of proof.

2 Transversals as Generating Sets

We have developed a technique that we call *shifting boxes*, which, for the sake of brevity, we will describe here as a systematic way to apply Nielsen transformations to a generating set of a group G, such that the resulting generators lie inside (or outside) particular desired cosets of a subgroup $H < G$. We cannot 'shift' generators in/out of any coset we like, but we do have a substantial degree of control.

Using this technique, we give the following equivalent conditions for a finite index subgroup to possess a left transversal which generates the whole group:

Theorem 7. *Let G be a finitely generated group and H a subgroup of finite index. Then the following are equivalent:*

1. *There exists a left transversal T for H in G with $\langle T \rangle = G$.*
2. *$|G : H| \geq d(G)$.*

In actual fact, we prove something much stronger: Given any generating set S for G and any subgroup H with $|G : H| \geq |S|$, there is a set S' Nielsen-equivalent to S (and thus $|S| = |S'|$) and an extension T of S' such that T is a left transversal for H in G.

It was pointed out to us by Chuck Miller that the above theorem follows from the Reidemeister–Schreier rewriting process. However, our techniques are much more elementary and also prove useful in obtaining other results that we were unable to obtain by direct application of Reidemeister–Schreier.

Recall that, if T is a left transversal for H in G, then T^{-1} (the set of inverses of all elements in T) is a right transversal of H in G. Moreover $\langle S \rangle = \langle S^{-1} \rangle$ for any subset S of a group G. So, as an immediate corollary, we see that the above result also carries over to right transversals.

The most natural question to ask now is 'When does a finite index subgroup have a left-right transversal which generates the whole group?' Combining our shifting boxes technique along with the properties of the coset intersection graph from Theorem 5, we are able to show the following:

Theorem 8. *Let G be a finitely generated group with $d(G) \leq 3$ and H a subgroup of finite index. Then the following are equivalent:*

1. *There exists a left-right transversal T for H in G with $\langle T \rangle = G$.*
2. *$|G : H| \geq d(G)$.*

Using Theorem 5, the proof of the above result is trivial when $d(G) = 1$, follows easily from 'shifting boxes' when $d(G) = 2$, and can be shown for the case $d(G) = 3$ via repeated application of 'shifting boxes'. Each case $d(G) = 1, 2, 3$ is shown by analysing an (increasing) finite number of possible scenarios. We have not yet extended this to the case $d(G) \geq 4$, as the number of scenarios to consider becomes very large and complex.

3 Applications to Primitive Elements in Free Groups

Recall that a *primitive* element of a finitely generated free group F_n is one which lies in *some* generating set of size precisely n.

Using existing techniques in group theory, we are able to show the following:

Theorem 9. *Let F_n be the free group on n generators and $N \lhd F_n$ a normal subgroup. Then the following are equivalent:*

1. *N contains some primitive element of F_n.*
2. *$d(F_n/N) < n$.*

Using our technique of shifting boxes, we obtain a related result:

Lemma 10. *If $H < F_n$ is a finite index subgroup with $|F_n : H| \leq n$, then every left (and hence every right) coset of H contains some primitive element of F_n.*

Our shifting boxes technique enables us to explore the location of primitive elements relative to cosets of a finite index subgroup H, in the following way:

Theorem 11. *Let H be a subgroup of finite index in F_n and suppose that the following inequality holds:*

$$|F_n : H| < n + \sum_{i=0}^{n} \binom{n}{i} \text{ (which is just } n + 2^n\text{)}.$$

Then at least one of the following occur:

1. *H contains at least one primitive element of F_n.*
2. *The square of every primitive element of F_n lies in H. When this occurs, H must be normal and contain $[F_n, F_n]$, so $F_n/H \cong C_2^m$ for some $m \leq n$.*

Moreover, if we reduce the bound to $|F_n : H| < 2^n$, then we may add to condition 2: 'every coset distinct from H contains a primitive element.'

In fact, the above theorem holds if we replace 'F_n' by 'a group G of rank n'. Of course, we must then define primitive element carefully; here we say it is an element in a generating set of G of size n.

By combining our results, we can give the following complete characterisation of finite index subgroups of F_n which contain primitive elements, up to index $n + 2^n$ (Theorem 9 shows $\Leftarrow$ and Theorem 11 shows $\Rightarrow$):

Theorem 12. *Let $H < F_n$ be of finite index with $|F_n : H| < n + 2^n$. Then*

$$H \text{ contains no primitive elements} \iff H \text{ is normal and } F_n/H \cong C_2^n.$$

Our analysis of the possible location of primitive elements relative to cosets of $H < F_n$ was motivated by the following result of Parzanchevski and Puder in [6]:

Theorem 13 ([6, Corollary 1.3]). *The set P of primitive elements in F_k is closed in the profinite topology.*

Corollary 14. *Given F_n and $w \in F_n$ a non-primitive element, there is a finite index subgroup $H < F_n$ such that the coset wH does not contain any primitive elements (but of course contains w). Taking $w = e$ gives a subgroup with no primitive elements.*

Given that the above result is existential, we decided to apply our techniques to look for explicit examples of subgroups with no primitive elements; Theorems 11 and 12 came about from this analysis, somewhat serendipitously.

4 Stallings Graphs

We note the following connection between any possible extension of Theorem 8 and free groups.

Lemma 15. *Theorem 8 is true for all finite rank groups if and only if it is true for all finite rank free groups.*

Thus we are able to restrict our investigation to the case of finite rank free groups. By doing this, we are able to make use of the theory of Stallings graphs for subgroups of free groups [3]. This simplification to free groups also extends to all earlier results in this work (they are true for all groups if and only if they are true for free groups). With the help of Enric Ventura and Jordi Delgado we have been able to give proofs to many of our results using the framework of Stallings graphs. However, attempts to generalise (or even re-prove) Theorem 8 using these techniques have so far been unsuccessful.

References

1. H. Chapman, A note on the elementary theory of groups of finite order. Messenger Math. **42**, 132–134 (1913)
2. H. Chapman, On a note on the elementary theory of groups of finite order. Messenger Math. **43**, 85 (1914)
3. I. Kapovich, A. Myasnikov, Stallings foldings and subgroups of free groups. J. Algebra **248**, 608–668 (2002)
4. G. Miller, On a method due to Galois. Q. J. Math. Oxf. Ser. **41**, 382–384 (1910)
5. O. Ore, On coset representatives in groups. Proc. Am. Math. Soc. **9**(4), 665–670 (1958)
6. O. Parzanchevski, D. Puder, Measure preserving words are primitive (2012), arXiv:1202.3269v1
7. G. Scorza, A proposito di un teorema del Chapman. Boll. Unione Math. Ital. **6**, 1–6 (1927)
8. S. Shü, On the common representative system of residue classes of infinite groups. J. Lond. Math. Soc. **16**, 101–104 (1941)
9. H. Zassenhaus, *Lehrbuch der Gruppentheorie* (German) (Teubner, Leipzig/Berlin, 1937)

Whitehead Problems for Words in $\mathbb{Z}^m \times F_n$

Jordi Delgado

We generically call *Whitehead problems* for a finitely presented group G the problems consisting in, given two objects (of the same certain suitable kind $\mathcal{O}$) in G, and a family $\mathcal{F}$ of transformations, decide whether there exists an element in $\mathcal{F}$ sending one object to the other. Specifically we will write WhP$(\mathcal{O}, \mathcal{F})$ to mean the Whitehead problem with objects in $\mathcal{O}$ and transformations in $\mathcal{F}$, i.e.,

$$\text{WhP}(\mathcal{O}, \mathcal{F}) \equiv \text{¿}\exists \varphi \in \mathcal{F} \text{ such that } o_1 \overset{\varphi}{\longmapsto} o_2 \text{?}_{(o_1, o_2 \text{ in } \mathcal{O})}.$$

It is customary to include as a part of the problem the search of one of such transformations, in case that it exists. So will we.

The kind of "objects in G" usually considered includes elements (i.e., words in the generators), subgroups, and conjugacy classes, as well as tuples of these, while the typical families of transformations (which we will always think of acting on the right) are those of automorphisms, monomorphisms, epimorphisms and endomorphisms of G. We denote them respectively by Aut G, Mon G, Epi G and End G.

This whole family of problems arise from the seminal WhP$(F_n, \text{Aut } F_n)$, where F_n denotes the free group on n generators. It was proposed and solved by Whitehead in [6] using a (now classical) technique called peak-reduction.

In this note we deal with Whitehead problems for words in finitely generated free-abelian times free groups (see [4] for full details). In sake of notational easiness we will hereafter usually abbreviate $G = \mathbb{Z}^m \times F_n$. Concretely we will solve WhP$(G, \text{Aut } G)$, WhP$(G, \text{Mon } G)$ and WhP$(G, \text{End } G)$. It is not surprising that the (already solved) corresponding problems for $\mathbb{Z}^m$ and F_n emerge when considering Whitehead problems for G.

J. Delgado (✉)
Departament de Matemàtica Aplicada III, Universitat Politècnica de Catalunya, Manresa, Spain
e-mail: jorge.delgado@upc.edu

J. González-Meneses et al. (eds.), *Extended Abstracts Fall 2012*,
Trends in Mathematics 1, DOI 10.1007/978-3-319-05488-9_6,
© Springer International Publishing Switzerland 2014

For the free-abelian groups the problems considered become those of the existence of solutions (of certain type) for integer matrix equations of the form $\mathbf{a} \cdot \mathbf{X} = \mathbf{b}$. This can be easily decided using linear algebra.

Proposition 1. *If $m \geq 1$, then*

(i) $\mathrm{WhP}(\mathbb{Z}^m, \mathrm{Aut}\, \mathbb{Z}^m)$ *is solvable;*
(ii) $\mathrm{WhP}(\mathbb{Z}^m, \mathrm{Mon}\, \mathbb{Z}^m)$ *is solvable;*
(iii) $\mathrm{WhP}(\mathbb{Z}^m, \mathrm{End}\, \mathbb{Z}^m)$ *is solvable.*   □

The same problems for the free group F_n are much more complicated. As mentioned above, the case of automorphisms was already solved by Whitehead back in the 1930s of the last century. The case of endomorphisms can be solved by writing a system of equations over F_n (with unknowns being the images of a given free basis for F_n), and then solving it by the powerful Makanin's algorithm. Finally, the case of monomorphisms was recently solved by Ciobanu and Houcine.

Theorem 2. *If $n \geq 2$, then*

(i) (Whitehead [6]) $\mathrm{WhP}(F_n, \mathrm{Aut}\, F_n)$ *is solvable;*
(ii) (Ciobanu–Houcine [1]) $\mathrm{WhP}(F_n, \mathrm{Mon}\, F_n)$ *is solvable;*
(iii) (Makanin [5]) $\mathrm{WhP}(F_n, \mathrm{End}\, F_n)$ *is solvable.*   □

So, the auto, mono and endo Whitehead problems (for words) are solvable for both $\mathbb{Z}^m$ and F_n. For $G = \mathbb{Z}^m \times F_n$ though, these problems turn out to be more than the mere juxtaposition of the corresponding problems for its factors. This is because the endomorphisms of G are more than pairs of endomorphisms of $\mathbb{Z}^m$ and F_n as well. It is not difficult to obtain a complete description of them imposing preservation of the (commutativity) relations defining G.

Proposition 3. *The endomorphisms of $G = \mathbb{Z}^m \times F_n$ are given by*

$$\Psi_{\phi, \mathbf{Q}, \mathbf{P}} \colon (\mathbf{a}, u) \longmapsto (\mathbf{a}\mathbf{Q} + \mathbf{u}\mathbf{P}, u\phi_{\mathbf{a}}),$$

where $\mathbf{u} = uab \in \mathbb{Z}^n$, $\mathbf{Q}$ and $\mathbf{P}$ are integer matrices, and $\phi_{\mathbf{a}} \colon F_n \to F_n$ is either

(i) *An endomorphism $\phi \colon F_n \to F_n$ (independent from $\mathbf{a}$), or*
(ii) *A map $u \mapsto w^{\alpha(\mathbf{a}, \mathbf{u})}$ where w is a non-proper power in $F_n \setminus \{1\}$ and $\alpha(\mathbf{a}, \mathbf{u}) = \mathbf{a}\mathbf{l}^\top + \mathbf{u}\mathbf{h}^\top \in \mathbb{Z}$ for certain $\mathbf{l} \in \mathbb{Z}^m \setminus \{\mathbf{0}\}$ and $\mathbf{h} \in \mathbb{Z}^n$.*

We will refer to them as type (I) *and* type (II) *endomorphisms of G respectively.*   □

Note that if $n = 0$ then type (I) and type (II) endomorphisms do coincide. Otherwise, it turns out that type (II) endomorphisms are a sort of degenerate case corresponding to a free contribution from the abelian part while all the injective and exhaustive endomorphisms of G are of type (I). Indeed, viewing $\mathbf{Q}$ as the endomorphism of $\mathbb{Z}^m$ given by right multiplying by $\mathbf{Q}$, we have the following quite natural characterization (note that the matrix $\mathbf{P}$ plays absolutely no role in this matter).

Proposition 4. *Let* Ψ *be an endomorphism of* $G = \mathbb{Z}^m \times F_n$, *with* $n \geq 2$. *Then*

(i) Ψ *is a monomorphism if and only if it is of type (I) with* ϕ *a monomorphism of* F_n *and* $\mathbf{Q}$ *a monomorphism of* $\mathbb{Z}^m$ *(i.e.,* $\det \mathbf{Q} \neq 0$);

(ii) Ψ *is an epimorphism if and only if it is of type (I) with* ϕ *an epimorphism of* F_n *and* $\mathbf{Q}$ *an epimorphism of* $\mathbb{Z}^m$ *(i.e.,* $\det \mathbf{Q} = \pm 1$).   □

The hopficity of $\mathbb{Z}^m$ and F_n together with this last proposition provide immediately the following results.

Corollary 5. $\mathbb{Z}^m \times F_n$ *is hopfian and not cohopfian.*   □

Corollary 6. *An endomorphism of* $G = \mathbb{Z}^m \times F_n$ $(n \geq 2)$ *is an automorphism if and only if it is of type (I) with* $\phi \in \mathrm{Aut}(F_n)$ *and* $\mathbf{Q} \in \mathrm{GL}_m(\mathbb{Z})$.   □

Now we have the ingredients to prove the main result of this note.

Theorem 7. *If* $G = \mathbb{Z}^m \times F_n$ *with* $m \geq 1$ *and* $n \geq 2$, *then*

(i) $\mathrm{WhP}(G, \mathrm{Aut}\, G)$ *is solvable;*

(ii) $\mathrm{WhP}(G, \mathrm{Mon}\, G)$ *is solvable;*

(iii) $\mathrm{WhP}(G, \mathrm{End}\, G)$ *is solvable.*

Sketch of the proof. We are given two elements $(\mathbf{a}, u)$, $(\mathbf{b}, v) \in G$, and have to decide whether there exists an automorphism (resp. monomorphism, endomorphism) of $\mathbb{Z}^m \times F_n$ sending one to the other; and in the affirmative case, find one of them.

Using the previous descriptions for each type of transformations in $\mathbb{Z}^m \times F_n$ and separating the free-abelian and free parts, our problems reduce to deciding whether there exist integer matrices $\mathbf{P}, \mathbf{Q}$ and a transformation ϕ of F_n ($\mathbf{Q}$ and ϕ of certain kind depending on the case; see Proposition 4) such that the two following independent conditions hold:

$$u\phi = v \left. \right\} \tag{1}$$
$$\mathbf{a}\mathbf{Q} + u\mathbf{P} = \mathbf{b} \left. \right\} . \tag{2}$$

Note that the subproblem associated to condition (1) becomes respectively the already solved $\mathrm{WhP}(F_n, \mathrm{Aut}\, F_n)$, $\mathrm{WhP}(F_n, \mathrm{Mon}\, F_n)$ and $\mathrm{WhP}(F_n, \mathrm{End}\, F_n)$ in the cases of autos, monos, and endos of type (I), and is straightforward to check for endos of type (II). Thus, if there is no ϕ solving each of these problems (for F_n), then the corresponding problem (for G) has no solution either, and we are done.

Otherwise, the decision method provides such a ϕ and our target reduces to solving the subproblem associated to condition (2): given arbitrary elements $\mathbf{a} \in \mathbb{Z}^m$ and $\mathbf{u} \in \mathbb{Z}^n$, decide whether there exist integer matrices $\mathbf{P}$ and $\mathbf{Q}$ (satisfying $\det \mathbf{Q} \neq 0$ in the case of monos, and $\det \mathbf{Q} = \pm 1$ in the case of autos) such that $\mathbf{a}\mathbf{Q} + \mathbf{u}\mathbf{P} = \mathbf{b}$.

If $\mathbf{a} = \mathbf{0}$ or $\mathbf{u} = \mathbf{0}$, these are well known results in linear algebra. Otherwise, write $\alpha = \gcd(\mathbf{a}) \neq 0$ and $\mu = \gcd(\mathbf{u}) \neq 0$. Then the problems reduce to test whether the linear system of equations

$$
\left.\begin{array}{c}
\alpha x_1 + \mu y_1 = b_1 \\
\vdots \\
\alpha x_m + \mu y_m = b_m
\end{array}\right\} \tag{3}
$$

has integral solutions $x_1, \ldots, x_m, y_1, \ldots, y_m \in \mathbb{Z}$ (with no extra condition in the case of endos, satisfying $(x_1, \ldots, x_m) \neq \mathbf{0}$ in the case of monos, and $\gcd(x_1, \ldots, x_m) = 1$ in the case of autos).

So, for the case of endos the decision is a standard argument in linear algebra. In the case of monomorphisms, the condition $(x_1, \ldots, x_m) \neq \mathbf{0}$ turns out to be superfluous and the same standard argument works, while the more involved case of autos becomes an exercise in elementary arithmetic and is decidable as well.

Finally, observe that in any of the affirmative cases we can use the description in Proposition 3 to reconstruct a transformation Ψ (of the corresponding type) such that $(\mathbf{a}, u)\Psi = (\mathbf{b}, v)$. □

We note that, very recently, a new version of the classical peak-reduction theorem has been developed by M. Day [3] for an arbitrary partially commutative group (see also [2]). These techniques allow the author to solve the Whitehead problem for this kind of groups, in its variant relative to tuples of conjugacy classes and automorphisms. As far as we know, WhP(G, Mon G) and WhP(G, End G) remain unsolved for a general partially commutative group G. Our Theorem 7 is a small contribution into this direction, solving these problems for free-abelian times free groups in a direct and self-contained form.

Acknowledgements The author thanks the hospitality of the Centre de Recerca Matemàtica (CRM Barcelona) along the research programme on Automorphisms of Free Groups during which this preprint was finished, and gratefully acknowledges support of Universitat Politècnica de Catalunya through PhD grant number 81–727 and partial support from the MEC (Spain) through project number MTM2011-25955.

References

1. L. Ciobanu, A. Houcine, The monomorphism problem in free groups. Arch. Math. (Basel) **94**(5), 423–434 (2010)
2. M.B. Day, Peak reduction and finite presentations for automorphism groups of right-angled Artin groups. Geom. Topol. **13**, 817–855 (2009)
3. M.B. Day, Full-featured peak reduction in right-angled Artin groups (2012), arXiv:1211.0078
4. J. Delgado, E. Ventura, Algorithmic problems for free-abelian times free groups. J. Algebra **391**, 256–283 (2013)
5. G. Makanin, Equations in free groups (Russian). Izv. Akad. Nauk SSSR Ser. Mat. **46**, 1190–1273 (1982)
6. J.H.C. Whitehead, On equivalent sets of elements in a free group. Ann. Math. **37**(4), 782–800 (1936)

Hausdorff Dimension and the Abelian Group Structure of Some Groups Acting on the p-Adic Tree

Gustavo A. Fernández-Alcober, Olivier Siegenthaler, and Amaia Zugadi-Reizabal

1 Introduction

Groups acting on rooted trees have gained interest since the 1980s, when the Grigorchuk groups [4] and the Gupta–Sidki [5] groups were described. Grigorchuk's groups were the first examples of groups with intermediate word growth, and they all have proved to satisfy unusual properties. The second Grigorchuk group and the Gupta–Sidki groups were generalized to a family of groups called GGS-groups. In this exposition we study the congruence quotients of GGS-groups, i.e. the quotients by the level stabilizers. In Sect. 2 we compute the orders of these quotients and as a consequence we get the Hausdorff dimension. On the other hand, in Sect. 3 we use algebraic geometry techniques to give a new description of the elements of the profinite closures of the so-called non-symmetric GGS-groups. This leads us to the discovery of an abelian group structure on these groups, where the Gupta–Sidki groups are included. The results presented in this extended abstract are published in [3] and [6] respectively.

If p is a prime and $X = \{1, \ldots, p\}$, the *p-adic tree* $\mathcal{T}$ is the tree whose set of vertices is the free monoid X^*, and the vertex v is descendant of the vertex u if $v = ux$ for some $x \in X$. The nth level is the set of all vertices of length n and if we consider only words of length $\leq n$, then we have a finite tree $\mathcal{T}_n$.

G.A. Fernández-Alcober (✉)
Matematika Saila, Euskal Herriko Unibertsitatea, Bilbao, Spain
e-mail: gustavo.fernandez@ehu.es

O. Siegenthaler
Departement Mathematik, ETH Zürich, Zürich, Switzerland
e-mail: olivier.siegenthaler@gmx.net

A. Zugadi-Reizabal
Matematikaren eta Zientzia Esperimentalen Didaktika Saila, Euskal Herriko Unibertsitatea, Bilbao, Spain
e-mail: amaia.zugadi@ehu.es

J. González-Meneses et al. (eds.), *Extended Abstracts Fall 2012*, Trends in Mathematics 1, DOI 10.1007/978-3-319-05488-9_7, © Springer International Publishing Switzerland 2014

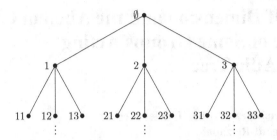

An *automorphism of* $\mathcal{T}$ is a bijection of the vertices that preserves incidence and we denote by $\operatorname{Aut}\mathcal{T}$ the group of all automorphisms of $\mathcal{T}$. The set $\operatorname{Stab}(n)$ of all automorphisms that fix the nth level is called the *nth level stabilizer*. These stabilizers are normal subgroups of $\operatorname{Aut}\mathcal{T}$ and can be considered as natural congruence subgroups for $\operatorname{Aut}\mathcal{T}$. We have $\operatorname{Aut}\mathcal{T}/\operatorname{Stab}(n) \cong \overbrace{S_p \wr (\cdots \wr (S_p \wr S_p))}^{n}$ and

$$\operatorname{Aut}\mathcal{T} = \varprojlim \operatorname{Aut}\mathcal{T}/\operatorname{Stab}(n) \cong \cdots \wr (S_p \wr (S_p \wr S_p))$$

is a profinite group.

If G is a subgroup of $\operatorname{Aut}\mathcal{T}$ and we put $\operatorname{Stab}_G(n) = \operatorname{Stab}(n) \cap G$, then we refer to the quotient $G_n = G/\operatorname{Stab}_G(n)$ as the *nth congruence quotient* of G. Since the kernel of the action of G on $\mathcal{T}_n$ is $\operatorname{Stab}_G(n)$, it follows that G_n can be naturally seen as a subgroup of $\operatorname{Aut}\mathcal{T}_n$.

On the other hand, any $f \in \operatorname{Aut}\mathcal{T}$ can be completely determined by giving, for every vertex u, the permutation of S_p that describes how f sends the descendants of u to the descendants of $f(u)$. More precisely, if $u \in X^*$, the permutation $\alpha \in S_p$ determined by $f(ux) = f(u)\alpha(x)$ for $x \in X$, is what we call the *label* f_u of f at u. And the collection of all labels $\{f_u\}_{u \in X^*}$ of f constitutes the *portrait* of f. The other way around, every labeling of the tree with permutations of S_p gives an automorphism of X^*.

An important automorphism of $\mathcal{T}$ is the automorphism a that permutes the main p subtrees of the tree rigidly according to the cycle $(1\ 2\ \dots\ p)$, i.e. the automorphism whose portrait is given by $a_u = 1$ for all u except from the root and $a_\emptyset = (1\ 2\ \dots\ p)$. Since a has order p, the expression a^e where $e \in \mathbb{F}_p$ makes sense. Now given a non-zero vector $\mathbf{e} = (e_1, \dots, e_{p-1}) \in \mathbb{F}_p$, we define the automorphism b by the following recurrence relation

$$b = (a^{e_1}, \dots, a^{e_{p-1}}, b).$$

We say that the subgroup $G = \langle a, b \rangle$ of $\operatorname{Aut}\mathcal{T}$ is the *GGS-group* corresponding to the *defining vector* $\mathbf{e}$. If $e_i = e_{p-i}$ for all $1 \le i \le (p-1)/2$, we say that the GGS-group is *symmetric*. Given a GGS-group corresponding to the vector $\mathbf{e}$, we may define the matrix C associated to $\mathbf{e}$ as the circulant matrix whose first row is

$(e_1, \ldots, e_{p-1}, 0)$. The rank $t = t(G)$ of that matrix will be said to be *the rank of the GGS-group.*

Throughout these pages p will denote an odd prime number.

2 Hausdorff Dimension

In this section we present the result published in [3], where we compute the orders of the congruence quotients of GGS-groups. As a consequence, we get the Hausdorff dimension of the profinite closures $\overline{G}$ of all GGS-groups.

Hausdorff dimension is a fractal dimension defined over metric spaces. It gives the relative size of a subset inside the big space.

Let G be a countably based profinite group i.e. there exists a descending chain $\{G(n)\}_{n \in \mathbb{N}}$ of open normal subgroups which form a base of neighbourhoods of the identity. If H is a closed subgroup of G, then, as Abercrombie [1] and Barnea and Shalev [2] proved, the Hausdorff dimension of H in G is

$$\dim_G H = \liminf_{n \to \infty} \frac{\log |HG(n)/G(n)|}{\log |G/G(n)|}.$$

Let us consider now the set Γ of all automorphisms of the tree that only have powers of the p-cycle $(1 \ldots p)$ in their portraits. Then Γ is a Sylow pro-p subgroup of Aut $\mathcal{T}$ and

$$\Gamma = \varprojlim \Gamma/\mathrm{Stab}_\Gamma(n) \cong \cdots \wr (C_p \wr (C_p \wr C_p)).$$

We compute Hausdorff dimension inside Γ with respect to the chain of the level stabilizers:

Theorem 1. *Let G be a GGS-group over the p-adic tree, where p is an odd prime, and let $\mathbf{e} \in \mathbb{F}_p^{p-1}$ be the defining vector of G. Then*

$$\log_p |G_n| = tp^{n-2} + 1 - r \frac{p^{n-2} - 1}{p - 1} - s \frac{p^{n-2} - (n-2)p + n - 3}{(p-1)^2}$$

for $n \geq 2$, and

$$\dim_\Gamma \overline{G} = \frac{(p-1)t}{p^2} - \frac{r}{p^2} - \frac{s}{p^2(p-1)},$$

where $t = t(G)$ is the rank of G,

$$r = \begin{cases} 1, & \text{if } \mathbf{e} \text{ is symmetric,} \\ 0, & \text{otherwise,} \end{cases} \quad \text{and} \quad s = \begin{cases} 1, & \text{if } \mathbf{e} \text{ is constant,} \\ 0, & \text{otherwise.} \end{cases}$$

3   Equations and Addition in Non-symmetric GGS-Groups

The results of this section are published in [6] and lead to the fact that non-symmetric GGS-groups admit an unexpected abelian group structure.

Let f be an automorphism in Γ. Since f_u is a power of the cycle $(1 \ldots p)$ for all $u \in X^*$, from this point onwards we will consider f_u as an element in $\mathbb{F}_p$, given by the value of the corresponding exponent.

Now, if we denote by $\mathbb{F}_p^{\mathcal{T}}$ the set of infinite sequences indexed by the vertices of $\mathcal{T}$, then the following map is a bijection

$$\begin{aligned} \Gamma &\longrightarrow \mathbb{F}_p^{\mathcal{T}} \\ f &\longmapsto \{f_u\}_{u \in \mathcal{T}}. \end{aligned}$$

From this point of view, two operations coexist in Γ: multiplication, given by $(fg)_u = f_u + g_{f(u)}$, and addition in $\mathbb{F}_p^{\mathcal{T}}$, hence in Γ, given by $(f + g)_u = f_u + g_u$.

Definition 2. Let $\mathcal{A}$ be the set of all continuous functions $\Gamma \to \mathbb{F}_p$.

It can be proven that $\mathcal{A}$ consists of polynomial functions on the variables

$$\begin{aligned} [u] \colon \Gamma &\longrightarrow \mathbb{F}_p \\ f &\longmapsto f_u, \end{aligned}$$

for $u \in X^*$.

If $u \in X^*$ and $P = P([u_1], \ldots, [u_k]) \in \mathcal{A}$ we define $u * P \in \mathcal{A}$ as $u * P = P([uu_1], \ldots, [uu_k])$.

Definition 3. Let $I \subseteq \mathcal{A}$ be an ideal and $S \subseteq I$. We say that S *generates I as a branching ideal* if $\{u * P \mid u \in X^* \text{ and } P \in S\}$ generates I as an ideal.

Definition 4. If V is a subset of Γ, we let $i(V) \subseteq \mathcal{A}$ denote the annihilator of V, i.e. the set of polynomial functions vanishing on V:

$$i(V) = \{P \in \mathcal{A} \mid P(f) = 0 \text{ for all } f \in V\}.$$

If I is a subset of $\mathcal{A}$, we let $\mathcal{V}(I)$ be the annihilator of I, i.e. the set

$$\mathcal{V}(I) = \{f \in \Gamma \mid P(f) = 0 \text{ for all } P \in I\}.$$

We will say that $f \in i(V)$ is an *equation* for V.

These two operators i and $\mathcal{V}$ behave as one may expect. See [6] for more information.

Theorem 5. *Let G be a GGS-group over the p-adic tree, where p is an odd prime. Let $\mathbf{e} \in \mathbb{F}_p^{p-1}$ be the defining vector of G and $t = t(G)$ the rank of G. Then there exist linear functions R_i and P_j, for $i = 1, \ldots, p - t$ and $j = 1, \ldots, p$, such that*

$$S = \{R_i, P_j \mid i = 1, \ldots, p - t, \ j = 1, \ldots, t\}$$

generates $\mathrm{i}(G)$ *as a branching ideal. As a consequence,*

$$\overline{G} = \mathcal{V}(\{v * Q \mid v \in X^*, \ Q \in S\}).$$

The linearity of these equations is crucial but not enough to prove the following theorem. Observe that if G is a non-symmetric GGS-group, Theorem 6 is straightforward from the preceding result for the profinite closure $\overline{G}$, but not for the abstract group G.

Theorem 6. *Let G be a non-symmetric GGS-group. Then the binary operation*

$$(f + g)_u = f_u + g_u,$$

for $f, g \in G$ and $u \in X^*$, gives G the structure of an abelian group.

Example 7. The Gupta–Sidki group, given by the vector $\mathbf{e} = (1, -1, 0, \ldots, 0) \in \mathbb{F}_p$, and the Fabrykowski–Gupta group, with $\mathbf{e} = (1, 0, 0)$, acquire the structure of abelian groups with respect to the pointwise addition described in the theorem.

Equations have proved to be useful for other purposes. For instance, one may recover Hausdorff dimension (observe that the values in Corollary 8 match with the ones in Theorem 1). By Theorem 2.21 from [6] we get

Corollary 8. *Let G be a non-symmetric GGS-group and $t = t(G)$ the rank of G. Then*

$$\dim_\Gamma \overline{G} = \frac{(p - 1)t}{p^2}.$$

On the other hand, there is work in progress suggesting that these algebraic geometry techniques might be very useful in decision problems as well.

References

1. A.G. Abercrombie, Subgroups and subrings of profinite rings. Math. Proc. Camb. Philos. Soc. **116**, 209–222 (1994)
2. Y. Barnea, A. Shalev, Hausdorff dimension, pro-p groups, and Kac-Moody algebras. Trans. Am. Math. Soc. **349**, 5073–5091 (1997)
3. G.A. Fernández-Alcober and A. Zugadi-Reizabal, GGS-groups: Order of congruence quotients and Hausdorff dimension. Trans. Amer. Math. Soc. **366** 1993–2017, (2014)
4. R.I. Grigorchuk, On Burnside's problem on periodic groups. Funktsional. Anal. i Prilozhen. **14**(1), 53–54 (1980) (Russian); English transl., Funct. Anal. Appl. **14**(1), 41–43 (1980)
5. N. Gupta, S. Sidki, On the Burnside problem for periodic groups. Math. Z. **182**(3), 385–388 (1983)
6. O. Siegenthaler, A. Zugadi-Reizabal, The equations satisfied by GGS-groups and the abelian group structure of the Gupta-Sidki group. Eur. J. Comb. **33**, 1672–1690 (2012)

Thompson's Group T, Undistorted Free Groups and Automorphisms of the Flip Graph

Ariadna Fossas

In this paper we present combinatorial and geometric results about Thompson's group T by comparing multiple viewpoints of the same object and their interactions (see [3] for an introduction to Thompson's groups). First, we find a non distorted subgroup of T isomorphic to the free non abelian group of rank 2 by using both the combinatorial version of T in terms of equivalence classes of tree pair diagrams, and Thurston's approach of T in terms of piecewise $PSL_2(\mathbb{Z})$ homeomorphisms of the real projective line (see [7, 10, 13]). Second, we study the action of T on a locally infinite graph $\mathcal{C}^1$ that can be seen as a generalisation of the flip graph (see, for example, [9]) for an infinitely sided convex polygon. The automorphism group of $\mathcal{C}^1$ is an extension of Thompson's group T.

1 An Undistorted F_2 in Thompson's Group T

The *piecewise linear Thompson's group* T is the group of orientation-preserving piecewise linear homeomorphisms of the circle S^1, viewed as the unit interval with identified endpoints, that map the set $\mathbb{Z}[1/2] \cap [0, 1]$ of dyadic rational numbers to itself and are differentiable except at finitely many points. We further assume that all non-differentiable points are dyadic rational numbers and, on intervals of differentiability, the derivatives are powers of 2.

A *tree pair diagram* is a triple (R, σ, S), where R and S are finite rooted binary trees with the same number n of leaves and σ is a bijection between the set of leaves of R and the set of leaves of S. The binary tree R is called the *source tree* and S is called the *target tree*. An internal vertex together with its two downward descendants is called a *caret*. A caret is *exposed* if it contains two leaves of the tree.

A. Fossas (✉)
Institut Fourier, Université Joseph Fourier, Grenoble, France
e-mail: ariadna.fossas@ujf-grenoble.fr

J. González-Meneses et al. (eds.), *Extended Abstracts Fall 2012*,
Trends in Mathematics 1, DOI 10.1007/978-3-319-05488-9__8,
© Springer International Publishing Switzerland 2014

A tree pair diagram is *reduced* if, for all exposed carets of S, the images under σ of their leaves do not form an exposed caret of R. A tree pair diagram is *cyclic* if σ is a cyclic permutation. In this case, the i-th leaf of the source tree (considering the left-to-right order on its leaves) is sent to the $\sigma(i)$-th leaf of the target tree. Two tree pair diagrams are *equivalent* if they have a common expansion. The *combinatorial Thompson's group* T is the set of equivalence classes of tree pair diagrams.

The *piecewise projective Thompson's group* T ($PPSL_2(\mathbb{Z})$ for short) is the group of orientation preserving homeomorphisms of the real projective line $\mathbb{R}P^1$ which are piecewise $PSL_2(\mathbb{Z})$ and have a finite number of non differentiable points, all of them being rational numbers. Using this version of T, it is natural to consider the subgroup of T which is isomorphic to the projective special linear group $PSL_2(\mathbb{Z})$. We will denote by a and b both the generators of

$$PSL_2(\mathbb{Z}) = \langle ab \mid a^2 = b^3 = 1 \rangle$$

as a group of matrices and their images under the isomorphism between $PPSL_2(\mathbb{Z})$ and Thompson's group T.

A finite rooted binary tree is *thin* if all its carets have one leaf and one internal vertex except for the last caret, which is exposed. It is easy to see that one can characterize thin trees with n leaves by associating a weight $r_i \in \{-1, 1\}$ to each one of its $n - 2$ internal vertices. Let $v_0, \ldots, v_{n-3}$ be the internal vertices of a thin tree. Denote by v_{-1} its root. Then, $r_i = 1$ if v_i is the left descendant of v_{i-1} and $r_i = -1$ if v_i is the right descendant of v_{i-1}.

Theorem 1. *Let (R, σ, S) be the reduced tree pair diagram of an element t of Thompson's group T. Then, t belongs to the subgroup $PSL_2(\mathbb{Z})$ if and only if (R, σ, S) satisfies one of the following conditions:*

(1) *The number of leaves of R and S is less than 4; or*
(2) *R is a thin tree with associated weights $r_0, \ldots, r_{k-1}$, S is a thin tree with associated weights $s_0, \ldots, s_{k-1}$, and the associated weights of R and S verify the equations*

$$\sum_{i=2}^{k-1} r_i s_{k+1-i} = 2 - k, \tag{1}$$

$$l + \sigma(1) + \epsilon(s_0) \equiv \frac{3 - s_1}{2} \pmod{k+2}, \tag{2}$$

where $l - 1$ is the cardinal of the set $\{i \ : \ r_i = -1, 0 \leq i \leq k - 1\}$, and

$$\epsilon(x) = \begin{cases} 1 \ \text{if } x = 1 \\ 0 \ \text{if } x = -1. \end{cases}$$

The proof of this result can be found in [4]. The following proposition proves the 'if' part of the theorem. For the 'only if' part, it suffices to show that the number of solutions to Eqs. (1) and (2) coincides with the number of words $w(a, b) = a^{\epsilon_1} b^{\delta_1} a b^{\delta_2} a \dots a b^{\delta_k} a^{\epsilon_2}$ with $\epsilon_1, \epsilon_2 \in \{0, 1\}$ and $\delta_1, \dots, \delta_k \in \{-1, 1\}$.

Proposition 2. *Let $w(a, b)$ be a word in normal form with respect to the generating system $\{a, b\}$ of $PSL_2(\mathbb{Z})$,*

$$w(a, b) = a^{\epsilon_1} b^{\delta_1} a b^{\delta_2} a \dots a b^{\delta_k} a^{\epsilon_2},$$

where $\epsilon_1, \epsilon_2 \in \{0, 1\}$ and $\delta_1, \dots, \delta_k \in \{-1, 1\}$. Assume that $k \geq 2$. Let R be the thin tree defined by the weights

$$r_0 = \epsilon^{-1}(\epsilon_1) \text{ and}$$
$$r_i = \delta_i, \ 1 \leq i \leq k - 1,$$

and let S be the thin tree defined by the weights

$$s_0 = \epsilon^{-1}(\epsilon_2) \text{ and}$$
$$s_i = -\delta_{k+1-i}, \ 1 \leq i \leq k - 1.$$

Let σ be the cyclic permutation defined by the equation

$$\sigma(1) \equiv \frac{3 - s_1}{2} - \epsilon(s_0) - l \pmod{k + 2},$$

where l and ϵ are defined as above.
      Then the reduced tree pair diagram of w is (R, σ, S), and the weights $r_0, \dots, r_{k-1}$ and $s_0, \dots, s_{k-1}$ satisfy Eqs. (1) and (2).

Using this results we can estimate the length of the elements in this subgroup. Furthermore, Burillo, Cleary, Stein and Taback gave an estimation of the length of elements in T in the number of carets in [2]. As a consequence of both results, we obtain:

Proposition 3. *The group $PSL_2(\mathbb{Z})$ is non distorted in Thompson's group T.*

Corollary 4. *Let $\{a, b\}$ be the standard generating set of $PSL_2(\mathbb{Z})$ as a subgroup of Thompson's group T. Then, the subgroup $H = \langle abab^{-1}, ab^{-1}ab \rangle$ is a free non abelian group of rank 2 and it is non distorted in T.*

2   An Infinite Graph Whose Automorphism Group is T

Let D denote the closed disk of $\mathbb{R}^2$ of center $(0, 0)$ and perimeter 1. Suppose its boundary S^1 arc-parametrized by the unit interval $[0, 1]$. The set of segments joining two different dyadic boundary points of D,

$$\Delta_F = \{(0, 1/2)\} \bigcup \left\{ \left(\frac{m}{2^n}, \frac{m+1}{2^n} \right) : m, n \in \mathbb{Z}_{>0}, \ m \in \{0, \ldots, 2^n - 1\}, \ n > 1 \right\},$$

is called the set of *Farey arcs* and it is a triangulation of D in the sense that $D - \Delta_F$ is a disjoint union of open triangles in the interior of D. This triangulation is minimal in the sense that the arcs do not cross and no proper subset of Δ_F defines a triangulation of D. The triangulation defined by Δ_F is called *Farey's dyadic triangulation* of D.

A *dyadic triangulation of Farey type* is a triangulation of the interior of D obtained from Δ_F by taking off a finite number of Farey arcs and replacing them by other non intersecting segments joining two different dyadic boundary points of D, such that the result is a minimal triangulation of D. Let $\mathcal{C}^1$ be the *flip graph*, which is defined as follows: the set of vertices coincides with the set of dyadic triangulations of Farey type, and two vertices are joined by an edge if and only if the associated dyadic triangulations of Farey type differ by a single arc. The flip graph can be equipped with the combinatorial metric.

Proposition 5. *The flip graph $\mathcal{C}^1$ is connected and it is not Gromov hyperbolic.*

The piecewise linear version of Thompson's group T shows that T acts on the set of rational dyadic numbers $\mathbb{Z}[1/2]$, thus it is easy to extend this action to the direct product $\mathbb{Z}[1/2] \times \mathbb{Z}[1/2]$. Moreover, every segment joining two dyadic points of the boundary of D is completely determined by its extremal points. This extended action of T can be used to define the action of T on $\mathcal{C}^1$. The image by $t \in T$ of a dyadic triangulation of Farey type is still of Farey type, since t has only a finite number of linear pieces and the derivatives are powers of two.

Proposition 6. *Thompson's group T acts on the flip graph $\mathcal{C}^1$ by automorphisms. Thus, we have a natural map $\Psi : T \to \mathrm{Aut}(C)$. Furthermore, Ψ is injective.*

The flip complex is optimal in the sense that its automorphism group is "essentially" T itself.

Theorem 7. *The automorphism group of the flip complex is isomorphic to the semi-direct product $T \rtimes \mathbb{Z}/2\mathbb{Z}$.*

Moreover, the co-kernel acts as a non trivial outer automorphism. Hence, we obtain a non trivial action of $\mathbb{Z}/2\mathbb{Z}$ on $\mathrm{Out}(T)$, which is an isomorphism by Brin's theorem (see [1]), so $\mathrm{Out}(T) \simeq \mathbb{Z}/2\mathbb{Z}$.

The 2-dimensional version of this result can be found in [5]. The proof of the theorem is based on the fact that every automorphism is determined by the image of the ball of radius one centered in some dyadic triangulation $\Delta \in \mathcal{C}^0$ of D of Farey type (when considering the combinatorial metric). Then, given an automorphism φ of the flip graph, we construct an element $t \in \Psi(T)$ such that $t(\Delta_F) = \varphi(\Delta_F)$. For this, one introduces an auxiliary complex, which is related to the link of Δ in a 2-dimensional version of $\mathcal{C}^1$.

The locally infinite graph C^1 can be generalised in all dimensions, as an infinite analogue to Stasheff's associahedra (see [12, 14, 15]). The first connection between Thompson's groups and Stasheff's associahedra was given by Greenberg in [8]. When considering the 2-dimensional version of the flip complex, the result about its automorphism group can be seen as an analogue to Ivanov's theorem about automorphisms of the curve complex (see [11]) using the modular version of Thompson's group T introduced by Funar, Kapoudjian and Sergiescu (see [6]).

Acknowledgements The author wishes to thank José Burillo, Louis Funar and Vlad Sergiescu for their comments and useful discussions. This work was partially supported by "Fundació La Caixa en partenariat avec l'Ambassade de France en Espagne", and "Fundación Caja Madrid".

References

1. M.G. Brin, The chameleon groups of Richard J. Thompson: automorphisms and dynamics. Publ. Math. IHÉS **84**, 5–33 (1996)
2. J. Burillo, S. Cleary, M. Stein, J. Taback, Combinatorial and metric properties of Thompson's group T. Trans. Am. Math. Soc. **361**, 631–652 (2009)
3. J.W. Cannon, W.J. Floyd, W.R. Parry, Introductory notes on Richard Thompson's groups. Enseign. Math. **42**, 215–256 (1996)
4. A. Fossas, $PSL_2(\mathbb{Z})$ as a non distorted subgroup of Thompson's group T. Indiana Univ. Math. J. **60**, 1905–1926 (2011)
5. A. Fossas, M. Nguyen, Thompson's group T is the orientation-preserving automorphism group of a cellular complex. Publ. Math. **56**(2), 305–326 (2012)
6. L. Funar, C. Kapoudjian, V. Sergiescu, Asymptotically rigid mapping class groups and Thompson's groups, in *Handbook of Teichmüller Theory*, vol. IV, ed. by A. Papadopoulos (European Mathematical Society Publishing House, Boston, 2012), pp. 595–664
7. E. Ghys, V. Sergiescu, Sur un groupe remarquable de difféomorphismes du cercle. Comment. Math. Helv. **62**, 185–239 (1987)
8. P. Greenberg, Les espaces de bracelets, les complexes de Stasheff et le groupe de Thompson. Bol. Soc. Mat. Mex. **37**, 189–201 (1992)
9. F. Hurtado, M. Noy, Graph of triangulations of a convex polygon and tree of triangulations. Comput. Geom. **13**(3), 179–188 (1999)
10. M. Imbert, Sur l'isomorphisme du groupe de Richard Thompson avec le groupe de Ptolémée, in *Geometric Galois Actions 2*. London Mathematical Society Lecture Note Series, vol. 243 (Cambridge University Press, Cambridge, 1997), pp. 313–324
11. N. Ivanov, Automorphism of complexes of curves and of Teichmüller spaces. Int. Math. Res. Not. **14**, 651–666 (1997)
12. C.W. Lee, The associahedron and triangulations of the n-gon. Eur. J. Comb. **10**, 551–560 (1989)
13. V. Sergiescu, Versions combinatoires de Diff(S^1), Groupes de Thompson, Prépublication de l'Institut Fourier no. 630 (2003)
14. J. Stasheff, Homotopy associativity of H-spaces I. Trans. Am. Math. Soc. **108**, 275–292 (1963)
15. J. Stasheff, Homotopy associativity of H-spaces II. Trans. Am. Math. Soc. **108**, 293–312 (1963)

Geometric Techniques in Braid Groups

Juan González-Meneses

The braid group with n strands B_n ($n \geq 1$) can be defined in several different ways:

- It is the fundamental group of the configuration space $\mathcal{C}_n$ of n distinct points in the plane, $B_n = \pi_1(\mathcal{C}_n)$.
- It is the mapping class group of the n-times punctured disk relative to the boundary of the disk, $B_n = \mathrm{MCG}(\mathbb{D}_n, \partial\mathbb{D}_n)$.
- It is the group with the following presentation, given by Artin [1,2]:

$$B_n = \left\langle \sigma_1, \ldots, \sigma_{n-1} \;\middle|\; \begin{array}{ll} \sigma_i\sigma_j = \sigma_j\sigma_i & |i - j| > 1 \\ \sigma_i\sigma_j\sigma_i = \sigma_j\sigma_i\sigma_j & |i - j| = 1 \end{array} \right\rangle.$$

From the algebraic point of view, one can consider the submonoid $B_n^+ \subset B_n$ generated (as a monoid) by $\sigma_1, \ldots, \sigma_{n-1}$. We define a partial order in B_n, called *prefix order*, saying that $a \preccurlyeq b$ if there exists $c \in B_n^+$ such that $ac = b$. This is a lattice order (every pair of elements $a, b \in B_n$ admits a unique gcd $a \wedge b$ and a unique lcm $a \vee b$), providing a structure of B_n called *Garside structure* which allows to define normal forms. Namely, if $\Delta = \sigma_1(\sigma_2\sigma_1)\cdots(\sigma_{n-1}\cdots\sigma_1)$, and we say that a braid α is *simple* if $1 \preccurlyeq \alpha \preccurlyeq \Delta$, then every braid x can be decomposed in a unique way as

$$x = \Delta^p x_1 \cdots x_r$$

where $p \in \mathbb{Z}$, x_i is a simple braid different from 1 and Δ, and $a_i^{-1}\Delta \wedge a_{i+1} = 1$ for $i = 1, \ldots, r - 1$.

J. González-Meneses (✉)
Departamento de Álgebra, Universidad de Sevilla, Sevilla, Spain
e-mail: meneses@us.es

J. González-Meneses et al. (eds.), *Extended Abstracts Fall 2012*,
Trends in Mathematics 1, DOI 10.1007/978-3-319-05488-9_9,
© Springer International Publishing Switzerland 2014

If we see braids as mapping classes, the element Δ^2 corresponds to a Dehn twist along a curve parallel to $\partial \mathbb{D}_n$. It is known that the center of B_n is infinite cyclic generated by Δ^2, and that $B_n/\langle \Delta^2 \rangle \simeq \text{MCG}(\mathbb{D}_n)$. One can then study braids using the Nielsen–Thurston theory of mapping classes. Then braids can be classified attending to their geometric nature, so that every braid is either *periodic*, or *reducible* (not periodic), or *pseudo-Anosov*.

Periodic braids correspond basically to rotations of the punctured disk, as they are conjugate either to a power of $\delta = \sigma_1 \cdots \sigma_{n-1}$ or to a power of $\varepsilon = \sigma_1 \delta$, which are conjugate, in the set of automorphisms of the punctured disk, to rigid rotations of angle $2\pi/n$ and $2\pi/(n-1)$, respectively.

Reducible braids are those which preserve a family of disjoint simple closed curves in $\mathbb{D}_n$, each one enclosing more than one and less than n punctures. Up to a conjugation, this means that they preserve a family of geometric circles centered at the diameter of $\mathbb{D}_n$. If one sees braids as loops in the configuration space, that is, as a motion of the n points in the disk, then this induces a motion of the invariant family of circles. Representing the graph of this motion, each circle defines a *tube*, so reducible braids can be seen as *tubular braids*, consisting of tubes and strands, each tube enclosing tubes and strands, and so forth.

Each invariant family of circles in the above sense is called a *reduction system* for the corresponding braid. For every reducible braid α there is a special invariant family of curves, $CRS(\alpha)$, called its *canonical reduction system* [5]. If a reduction system is a family of geometric circles we say that it is *standard*.

One connection between the algebraic and the geometric points of view was given in [10] where, for every reducible braid $a \in B_n$, its *standardizer*

$$St(a) = \{b \in B_n^+ \mid CRS(b^{-1}ab) \text{ is standard}\}$$

is shown to be closed under $\wedge$. This implies that there exists a unique minimal element for $\preccurlyeq$ in $St(a)$. In other words, given a reducible braid a, there is a unique minimal positive braid which conjugates a to a tubular braid.

We are then looking at the action of B_n on the set of disjoint simple essential curves in $\mathbb{D}_n$ (the *complex of curves*). Notice that periodic braids are those having a power which preserve *all* such families of curves, that is, they have a power whose induced action is trivial (their action is periodic). Reducible braids are those preserving some family of curves, hence their action has some fixed point. We remark that a periodic braid can also be reducible, so the word *reducible* sometimes refers to those reducible braids which are not periodic.

Finally, if the action induced by a braid α is not periodic and has no fixed point, the braid is said to be *pseudo-Anosov*. In this case, for every essential curve $\mathcal{C}$, its image under α^m as m goes to infinity tends to a foliation $\mathcal{F}_s$ of $\mathbb{D}_n$ with singular points and admitting a transverse measure. The image of α^{-m} thus tends to another transverse measured foliation $\mathcal{F}_u$. The action of α on $\mathcal{F}_u$ (called unstable foliation) scales its measure by some real number $\lambda > 1$, while the action on $\mathcal{F}_s$ (called stable foliation) scales its measure by $1/\lambda$ [11].

This geometric characterization of braids can be used to show some algebraic properties of the elements of B_n, for instance that roots of a braid (of the same index) are always conjugate, a former conjecture by Makanin.

Theorem 1 ([7]). *Given $a, b \in B_n$, if there is some $k > 0$ such that $a^k = b^k$ then a and b are conjugate.*

Sketch of proof. The braids a and b have the same geometric type, as this is preserved by taking powers.

If a and b are periodic, just by looking at their exponent sum as words in Artin's generators, one shows easily that they are conjugate to the same power of either δ or ε, hence conjugate to each other.

If a and b are pseudo-Anosov, then they preserve the same measured foliations and have the same dilatation factor λ, as the foliations are preserved by taking powers, and the dilatation factor of a^k is λ^k if and only if the dilatation factor of a is λ. This implies that the commutator $[a, b] = aba^{-1}b^{-1}$ preserves the same two foliations and has dilatation factor 1, therefore it is periodic. As its exponent sum is zero, $[a, b]$ is trivial, hence a and b commute. Therefore

$$a^k = b^k \Rightarrow a^k b^{-k} = 1 \Rightarrow (ab^{-1})^k = 1,$$

hence, as B_n has no torsion, $ab^{-1} = 1$, that is, $a = b$. Notice that in the pseudo-Anosov case we have actually shown that roots are unique.

If a and b are reducible, they have the same canonical reduction system, as this is preserved by taking powers. Up to conjugacy we can assume that they preserve the same family of circles. Their external *tubular braids* must then be conjugate by the previous two cases, as they are either periodic or pseudo-Anosov (otherwise they would not be the outermost tubes). Then, looking carefully at the strands inside the tubes, conjugating in the appropriate way and using induction on the number of strands, the result follows. See [7] for details. □

Another interesting application of the geometric properties of braids concerns centralizers of elements. Indeed, the structure of the centralizer of a braid heavily depends on its geometric type.

1. If $a \in B_n$ is pseudo-Anosov, then $Z(a) \simeq \mathbb{Z}^2$, freely generated by a pseudo-Anosov and by a periodic braid which commute (McCarthy JD, Normalizers and centralizers of pseudo-Anosov mapping classes, www.math.msu.edu/~mccarthy/publications/normcent.pdf, unpublished).
2. If $a \in B_n$ is periodic, then $Z(a)$ consists of *symmetric braids*, that is, braids which are preserved by the rotation induced by a. In this case, $Z(a)$ is isomorphic to the braid group of an annulus (defined by replacing the punctured disc by a punctured annulus) [4].
3. If $a \in B_n$ is reducible, then one can define its tubular or *external braid* a_{ext}, and its *internal braids* $a_1, \ldots, a_t$, and one has

$$Z(a) \simeq (Z(a_1) \times \cdots \times Z(a_t)) \rtimes Z_0(a_{ext}),$$

where $Z_0(a_{ext})$ is the subgroup of $Z(a_{ext})$ consisting of those braids inducing the appropriate permutation on the tubes [8].

Finally, the geometric properties of braids can be used to improve the usual solution to the conjugacy problem in B_n.

Usually, one solves the conjugacy problem in B_n by computing the set of *simplest* conjugates of a braid. Given $a \in B_n$, we can apply to it a special conjugation called *cyclic sliding*, depending only on the Garside normal form of a [6]. This conjugation never increases the number of factors in a normal form, hence iterated application of cyclic sliding always reaches a repetition. The *set of sliding circuits* of a, $SC(a)$, is the set of braids in the conjugacy class of a which belong to a periodic orbit under cyclic sliding. This is an invariant of the conjugacy class, hence a and b are conjugate if and only if $SC(a) = SC(b)$. Therefore, computing these sets one solves the conjugacy problem in B_n [8].

Basically, cyclic slidings simplify braids from the algebraic point of view, so $SC(a)$ is the set of "simplest" conjugates of a in some algebraic sense. The interesting property is that cyclic sliding also simplifies braids geometrically.

It follows from [3] that if a braid a is reducible, some element in $SC(a)$ has a standard canonical reduction system (its reducing curves are geometric circles, not tangled curves). Furthermore, if one defines $SC^*(a)$ to be the set of conjugates of a all of whose powers belong to periodic orbits under cyclic sliding (a set that can be effectively computed [9]), then *all* elements in $SC^*(a)$ admit a reduction curve which is either a geometric circle, or *almost* (where the definition of being *almost a circle* is clearly defined in [9]).

In the case of pseudo-Anosov braids, one cannot expect to simplify the invariant measured foliations by a conjugation. But we conjecture that applying cyclic sliding to a pseudo-Anosov braid simplifies the train tracks which are used to describe its invariant measured foliations. Hence one should be able to find geometrically simple train tracks for every element in $SC(a)$, for any given pseudo-Anosov braid a.

Acknowledgements Partially supported by MTM2010-19355, P09-FQM-5112, FEDER and the Australian Research Council's Discovery Projects funding scheme (project number DP1094072).

References

1. E. Artin, Theorie der Zöpfe. Abh. Math. Semin. Univ. Hambg. **4**, 47–72 (1925)
2. E. Artin, Theory of braids. Ann. Math. **48**(2), 101–126 (1947)
3. D. Benardete, M. Gutierrez, Z. Nitecki, Braids and the Nielsen–Thurston classification. J. Knot Theory Ramif. **4**(4), 549–618 (1995)
4. D. Bessis, F. Digne, J. Michel, Springer theory in braid groups and the Birman-Ko-Lee monoid. Pac. J. Math. **205**(2), 287–309 (2002)
5. J.S. Birman, A. Lubotzky, J. McCarthy, Abelian and solvable subgroups of the mapping class groups. Duke Math. J. **50**(4), 1107–1120 (1983)
6. V. Gebhardt, J. González-Meneses, The cyclic sliding operation in Garside groups. Math. Z. **265**(1), 85–114 (2010)

7. J. González-Meneses, The n-th root of a braid is unique up to conjugacy. Algebr. Geom. Topol. **3**, 1103–1118 (2003)
8. J. González-Meneses, B. Wiest, On the structure of the centralizer of a braid. Ann. Sci. Éc. Norm. Sup. **37**(5), 729–757 (2004)
9. J. González-Meneses, B. Wiest, Reducible braids and Garside theory. Algebr. Geom. Topol. **11**, 2971–3010 (2011)
10. S.J. Lee, E.K. Lee, A Garside-theoretic approach to the reducibility problem in braid groups. J. Algebra **320**(2), 783–820 (2008)
11. W.P. Thurston, On the geometry and dynamics of diffeomorphisms of surfaces. Bull. Am. Math. Soc. (N.S.) **19**(2), 417–431 (1988)

Induced Automorphisms

Vincent Guirardel and Gilbert Levitt

Rather than looking for subgroups invariant (up to conjugacy) under a given automorphism of a group G, one can fix a subgroup $H \subset G$ and consider automorphisms leaving H invariant.

Define $\operatorname{Out}(H \upharpoonright G) \subset \operatorname{Out}(H)$ as the group of outer automorphisms of H which extend to G. A general question (that we heard from D. Calegari) is to understand $\operatorname{Out}(H \upharpoonright G)$: What does it contain? Is it finitely generated? Is it finitely presented?...

If H is a vertex group G_v in a graph of groups decomposition of G, it is easy to see that $\operatorname{Out}(H \upharpoonright G)$ contains $\operatorname{Out}(H; \operatorname{Inc}_v)$, the group of automorphisms of H which act trivially (i.e., as an inner automorphism of H) on each incident edge group.

When H is a malnormal subgroup of F_n, this construction accounts for almost all of $\operatorname{Out}(H \upharpoonright G)$:

Theorem 1. *Let H be a finitely generated malnormal subgroup of F_n. If $\operatorname{Out}(H \upharpoonright F_n)$ is infinite, then H is a vertex group in a graph of groups decomposition of G with finitely generated edge groups. Moreover, $\operatorname{Out}(H; \operatorname{Inc}_v)$ has finite index in $\operatorname{Out}(H \upharpoonright F_n)$.*

It follows that $\operatorname{Out}(H \upharpoonright F_n)$ is finitely presented (it has a finite index subgroup with a finite classifying space).

The theorem also holds when H is a malnormal quasiconvex subgroup of a hyperbolic group G.

V. Guirardel (✉)
Institut de Recherche en Mathématiques de Rennes, Université de Rennes 1, Rennes, France
e-mail: vincent.guirardel@univ-rennes1.fr

G. Levitt
Laboratoire de Mathématiques Nicolas Oresme, Université de Caen et CNRS (URM 6139), Caen, France
e-mail: levitt@unicaen.fr

J. González-Meneses et al. (eds.), *Extended Abstracts Fall 2012*,
Trends in Mathematics 1, DOI 10.1007/978-3-319-05488-9_10,
© Springer International Publishing Switzerland 2014

For the proof, one may assume that H is not contained in a proper free factor. Malnormality ensures that F_n is hyperbolic relative to H (Bowditch). There is a preferred JSJ tree relative to H over subgroups which are cyclic or parabolic (contained in conjugates of H). Vertex groups are conjugate to H, or rigid, or surface groups (quadratically hanging subgroups). $\mathrm{Out}(H \upharpoonright F_n)$ infinite implies that H cannot be contained in a rigid vertex group (by the Bestvina–Paulin method, extended to relatively hyperbolic groups by Belegradek–Szczepański, and Rips theory of groups acting on $\mathbb{R}$-trees). Since H is not cyclic, it cannot be contained in a surface group, so H is a vertex group G_v. For the moreover, the action of $\mathrm{Out}(H \upharpoonright F_n)$ on an incident edge group G_e is controlled because the endpoint w other than v is a surface vertex (in which case $G_e = \mathbb{Z}$ and $\mathrm{Out}(G_e)$ is finite) or is rigid (in which case there are only finitely many possible actions on G_w).

The theorem easily implies:

Corollary 2. *If $H \subset F_n$ is finitely generated and malnormal, but not a free factor, then $\mathrm{Out}(H \upharpoonright F_n)$ contains no iwip (irreducible with irreducible powers) automorphism of H.*

The theorem does not hold if H is not malnormal, but the corollary extends:

Theorem 3. *Let $H \subset F_n$ be finitely generated. If $\mathrm{Out}(H \upharpoonright F_n)$ contains an iwip automorphism Φ, then H has finite index in a free factor.*

Corollary 4. *If H is not cyclic and $\mathrm{Out}(H \upharpoonright F_n) = \mathrm{Out}(H)$, then H is a free factor.*

These results also have extensions to the case when G is a hyperbolic group.

To prove Theorem 3, one may assume:

1. H is not malnormal;
2. H is not contained in a proper free factor;
3. There is no subgroup $H' \neq H$ containing H with finite index.

To achieve (3), one replaces H by its commensurator. Under these assumptions, one proves $H = F_n$.

By (1), there are non-trivial subgroups $H \cap H^g$ with $g \notin H$, and by (3) they have infinite index in H. Up to conjugacy in H, there are only finitely many groups $H \cap H^g$ (for G hyperbolic, this is due to Gitik–Mitra–Rips–Sageev). Some power of Φ leaves these groups invariant. By a result due to Bestvina–Feighn–Handel, it follows that Φ is geometric: H may be identified with the fundamental group of a surface Σ with one boundary component C, and Φ is induced by a pseudo-Anosov map f.

Since the cyclic group $A = \pi_1(C)$ cannot be contained in a proper free factor of F_n by (2), we can consider the preferred cyclic JSJ tree T of F_n relative to A. Note that H is elliptic in T: otherwise one would get a finite f-invariant collection of curves on Φ. The only possibility now is that H is contained in a surface vertex group G_v, necessarily with finite index. Finally, (2) and (3) imply that $H = F_n$.

Dual Automorphisms of Free Groups

Fedaa Ibrahim and Martin Lustig

Throughout this extended abstract of our joint work [2] we denote by F_N the non-abelian free group of finite rank $N \geq 2$. We assume familiarity of the reader with the basic terminology around the combinatorial theory of automorphisms of free groups.

In order to work with free groups, in particular for algorithmic purposes, one works almost always with a fixed basis $\mathcal{A}$ of F_N. In this case one has a canonical bijection between elements of F_N and the set $F(\mathcal{A})$ of reduced word in $\mathcal{A} \cup \mathcal{A}^{-1}$.

The distinction between reduced words and free group elements, though unusual in the classical approach to free groups, is rather important for the work presented here. The reason is that a reduced word $w = y_1 \cdots y_r \in F(\mathcal{A})$ allows a second, *dual* interpretation: Every such w determines a subset of ∂F_N, defined by the *cylinder* $C_w^1 \subset \partial F(\mathcal{A})$, where by $\partial F(\mathcal{A})$ we denote the set of infinite reduced words in $\mathcal{A} \cup \mathcal{A}^{-1}$:

$$C_w^1 = \{x_1 x_2 \cdots \mid x_1 = y_1, \ldots, x_r = y_r\}.$$

Note that the subset of ∂F_N defined by C_w^1 depends not only on the element of F_N given by the word w, but also on the chosen basis $\mathcal{A}$. An equivalent way to express this dependency is to note that for any automorphism $\phi \in \mathrm{Aut}(F_N)$ one has, in general,

$$\phi(C_w^1) \neq C_{\phi(w)}^1.$$

Indeed, the image of a cylinder under an automorphism $\phi \in \mathrm{Aut}(F_N)$ is in general not a cylinder, but a multi-cylinder, i.e., a finite union of cylinders. In his thesis and a subsequent publication (see [1]), the first author of the work presented here has

F. Ibrahim (✉) • M. Lustig
LATP, Centre de Mathématiques et Informatique, Aix-Marseille Université, Marseille, France
e-mail: fidaa0@hotmail.fr; Martin.Lustig@univ-amu.fr

J. González-Meneses et al. (eds.), *Extended Abstracts Fall 2012*,
Trends in Mathematics 1, DOI 10.1007/978-3-319-05488-9_11,
© Springer International Publishing Switzerland 2014

given an efficient algorithm how to determine this finite union, and in particular he has proved the following formula:

Theorem 1 ([1]).

(a) *Let φ be an automorphism of the free group F_N with finite basis $\mathcal{A}$. For any $u \in F(\mathcal{A})$ there exists a finite set $U \subset F(\mathcal{A})$ such that*

$$\varphi(C_u^1) = C_U^1 := \bigcup_{u_i \in U} C_{u_i}^1.$$

(b) *A set U as in statement (a) can be algorithmically derived from $u \in F(\mathcal{A})$ and from the words in the finite subsets $\varphi(\mathcal{A})$ and $\varphi^{-1}(\mathcal{A})$ of $F(\mathcal{A})$. Indeed, the equality in (a) is true for*

$$U = \{\varphi(u')|_{S(\varphi)^2} \mid u' \in u|^k\},$$

with $k = S(\varphi)^4 + S(\varphi)^3 + S(\varphi)^2$, where $S(\varphi)$ is the maximal length of any $\varphi(a_i)$ or $\varphi^{-1}(a_i)$ among all $a_i \in \mathcal{A}$.

Here for any reduced word $w \in F(\mathcal{A})$ and any integer $l \geq 0$ we denote by $w|_l$ the word obtained from u by erasing the last l letters, and by $w|^l$ the set of all reduced words obtained from u by adding l letters from $\mathcal{A} \cup \mathcal{A}^{-1}$ at the end of w.

A multi-cylinder $C_U^1 \subset \partial F(\mathcal{A})$ does not uniquely define the finite set $U \subset F(\mathcal{A})$, but in [1] it has been shown that there is a unique *reduced subset* $U_{\min} \subset F(\mathcal{A})$ of minimal cardinality which satisfies $C_U^1 = C_{U_{\min}}^1$, and that $U_{\min}$ can be derived from U by an elementary reduction algorithm. We use these facts to define for any $\phi \in \text{Aut}(F_N)$ the *dual automorphism* $\phi_{\mathcal{A}}^*$ as follows:

Definition 2. For any $u \in F(\mathcal{A})$, let $\phi_{\mathcal{A}}^*(u)$ be the reduced subset of $F(\mathcal{A})$ that satisfies

$$C_{\phi_{\mathcal{A}}^*(u)}^1 = \phi(C_u^1).$$

The reader should be warned that, as is indicated by the name, the dual automorphism depends heavily on the choice of the basis $\mathcal{A}$ of F_N.

There are several immediate natural questions which come to mind if one considers the definition of $\phi_{\mathcal{A}}^*$, concerning for example the size of the set $\phi_{\mathcal{A}}^*(u) \subset F_N$, its computability, and its behavior under iteration of $\phi_{\mathcal{A}}^*$. We will state now the main results of our joint work [2]; in particular, Theorem 5 seems surprising and noteworthy to us. The following result is inspired by Theorem 1 above and proved by similar methods:

Theorem 3. *For any automorphism $\phi \in \text{Aut}(F_N)$ and any basis $\mathcal{A}$ of F_N there exists a finite collection $\mathcal{U}(\phi_{\mathcal{A}}^*) = \{U_1, \ldots, U_r\}$ of finite sets $U_i \subset F(\mathcal{A})$ of reduced words, such that for every $w \in F(\mathcal{A})$ the dual image $\phi_{\mathcal{A}}^*(w)$ is given by vU_i, for*

some $v \in F(\mathcal{A})$ and $U_i = U_i(w) \in \mathcal{U}(\phi_{\mathcal{A}}^*)$. The word v can be specified further to
$v = \phi(w|_k)|_K$ for some constants $k, K \geq 0$ which are independent of w.

In particular, the cardinality of $\phi_{\mathcal{A}}^*(w)$, for any $w \in F(\mathcal{A})$, is bounded above by
a constant which only depends on ϕ (and the fixed basis $\mathcal{A}$).

The computation of the constants k and K and of the finite sets U_i is possible by the use of Theorem 1, and writing an efficient computer program should not be a very difficult task.

However, the result stated below in Theorem 5 is much more striking and also more useful for computational purposes; it came about when we tried to look at the easiest non-trivial example, namely an elementary Nielsen automorphism:

Remark 4 (Fundamental formulas). We recall Nielsen's famous result that every automorphism of a free group can be written as product of *elementary automorphisms*, which can be regrouped into the following two types:

(a) For the "standard", the elementary Nielsen automorphism $\phi \in F(\mathcal{A})$, for $\mathcal{A} = \{a, b, c_1, \ldots, c_q\}$, given by $a \mapsto ab, b \mapsto b, c_j \mapsto c_j$, we obtain (where we use the convention that a product of words uv is written as $u \cdot v$ if no cancellation occurs between the end of u and the beginning of v):

$$\phi_{\mathcal{A}}^*(w \cdot a) = \{\phi(w) \cdot a\}$$
$$\phi_{\mathcal{A}}^*(w \cdot a^{-1}) = \{\phi(w)b^{-1} \cdot a^{-1}\}$$
$$\phi_{\mathcal{A}}^*(w \cdot b) = \{\phi(w) \cdot b, \phi(w) \cdot a^{-1}\}$$
$$\phi_{\mathcal{A}}^*(w \cdot b^{-1}) = \{\phi(w)b^{-1} \cdot b^{-1}, \phi(w)b^{-1} \cdot a\}$$
$$\phi_{\mathcal{A}}^*(w \cdot c_j) = \{\phi(w) \cdot c_j\}$$
$$\phi_{\mathcal{A}}^*(w \cdot c_j^{-1}) = \{\phi(w) \cdot c_j^{-1}\}.$$

(b) The other class of elementary automorphisms ϕ is given by a permutation of the letters in $\mathcal{A}$, or by an inversion of some of them. In other words, one has $|\phi(a_i)| = 1$ for every $a_i \in \mathcal{A}$, and there is never a cancellation in the image of a reduced word. As a consequence, one obtains directly

$$\phi_{\mathcal{A}}^*(w) = \{\phi(w)\}$$

for any $w \in F(\mathcal{A})$.

These fundamental formulas give a straightforward method to calculate the analogous formulas for any automorphism, by applying the above fundamental formulas in an iterative way. This leads directly to:

Theorem 5. *For any automorphism $\phi \in \mathrm{Aut}(F(\mathcal{A}))$ the collection $\mathcal{U}(\phi_{\mathcal{A}}^*)$ from Theorem 3 can be specified to consist of precisely $2N$ sets $U(x)$, with $x \in \mathcal{A} \cup \mathcal{A}^{-1}$, such that for every reduced word $w = y_1 \cdots y_q$ one has (using the terminology of Theorem 3):*

$$U_i(w) = U(y_q).$$

Furthermore, if ϕ is the product of elementary automorphisms (i.e., basis permutations, basis inversions, or elementary Nielsen automorphisms), then the cardinality of each $U(x)$ is bounded by 2^t, where t is the number of elementary Nielsen automorphisms in the above decomposition of ϕ, and the constants $k \geq 0$ and $K \geq 0$ from Theorem 3 can be chosen as $k = 1$ and $K = 0$.

The 2^t-bound from the last theorem suggests the definition of a *dual growth rate* λ_ϕ^* which can be defined for example as the limit superior of the family of values $(\mathrm{card}(\phi^k)_{\mathcal{A}}^*(x))^{\frac{1}{k}}$, for $x \in \mathcal{A} \cup \mathcal{A}^{-1}$ and $k \in \mathbb{N}$.

Theorem 6. *For any automorphism $\phi \in \mathrm{Aut}(F_N)$ the dual growth rate λ_ϕ^* is independent of the choice of the basis $\mathcal{A}$. It can be calculated as the Perron–Frobenius eigenvalue of a $2N \times 2N$-matrix which can be algorithmically derived from ϕ.*

It turns out that cylinders are less natural objects in group theory than in combinatorics. For example, one has $vC_u^1 = C_{vu}^1$ only if u is not completely cancelled when reducing vu. This (and related) problems vanish if one passes over to *double cylinders* $C_{[v,w]}^2$, for $v, w \in F(\mathcal{A})$, which are defined as sets of endpoint pairs of biinfinite reduced paths in the Cayley tree $\tilde{\Gamma}_{\mathcal{A}}(F_N)$ that pass through both vertices v and w. All the results stated above for single cylinders have natural analogues for double cylinders (compare with [1]). Double cylinders play a natural role in defining currents for free groups and in properly setting up the basic theory of such; see [3, 4]. Indeed, one way to name concretely a current μ is to specify its Kolmogorov function, which is the non-negative function $\mu_{\mathcal{A}} : F(\mathcal{A}) \to \mathbb{R}$ which associates to every $w \in F(\mathcal{A})$ the current measure $\mu(C_{[1,w]}^2)$ of the double cylinder $C_{[1,w]}^2 \subset F_N \times F_N \smallsetminus \mathrm{diagonal}$.

In §6 of [4] a formula is obtained to derive from the Kolmogorov function for a current μ the one for the image current $\phi(\mu)$, for any $\phi \in \mathrm{Aut}(F_N)$. The above finiteness statements and the algorithmic applications allow a substantial simplification and specification of this formula. The interest towards a concrete and effective calculation of $\phi(\mu)$ from known data for ϕ and μ was one of the original motivations to stimulate this work.

References

1. F. Ibrahim, Cylinders, multi-cylinders and the induced action of $\mathrm{Aut}(F_n)$. Groups Complex. Cryptol. **4**, 357–375 (2012)
2. F. Ibrahim, M. Lustig, Dual automorphisms of free groups preprint, available at http://arxiv.org/pdf/1306.5696v1.pdf (2013)
3. I. Kapovich, The frequency space of a free group. Int. J. Algebra Comput. **15**, 939–969 (2005)
4. I. Kapovich, Currents on free groups, in *Topological and Asymptotic Aspects of Group Theory*, ed. by R. Grigorchuk, M. Mihalik, M. Sapir, Z. Sunik. Contemporary Mathematics, vol. 394 (American Mathematical Socity, Providence, 2006), pp. 149–176

Effective Algebraic Detection of the Nielsen–Thurston Classification of Mapping Classes

Thomas Koberda and Johanna Mangahas

We describe a "list-and-check" algorithm for determining the Nielsen–Thurston type of elements of the mapping class group of a surface, that is, whether $f \in \mathrm{Mod}(S)$ is finite-order, pseudo-Anosov, or reducible. The algorithm also detects the *canonical reduction system* $\sigma(f)$ of f, which is the unique set of disjoint curves on S fixed setwise by f, common to all maximal such sets. The algorithm has a simple underlying idea: compute an upper bound for the length of curves in $\sigma(f)$, list all sufficiently short curves, and check their images under an appropriate power of f.

Bestvina and Handel gave the first $\mathrm{Mod}(S)$ classification algorithm, which determines the train track for the limiting lamination of a mapping class group element [3]. While this algorithm is fast in practice, its complexity is unknown. Chen and Hamidi-Tehrani obtained an exponential-time classification using the piecewise-linear $\mathrm{Mod}(S)$-action on measured train-tracks [7]. Braid groups are closely related to mapping class groups of punctured disks and are likewise classified. For these, Calvez has recently shown the existence of a polynomial time classification algorithm [4]; see also [1,2,5,6,9]. Using a polynomial time algorithm to enumerate simple curves of bounded length, our classification algorithm is exponential in the word length of f.

Let S be a surface with genus g and p punctures, and negative Euler characteristic $\chi(S) = 2 - 2g - p$. Fix a generating set X for $\mathrm{Mod}(S)$, and, for $f \in \mathrm{Mod}(S)$, let $|f|$ denote word length with respect to X. Fix a definition of length $\ell(\cdot)$ for curves on S from the array of reasonable options (e.g., choice of metric on S, word metric

T. Koberda (✉)
Department of Mathematics, Yale University, New Haven, CT, USA
e-mail: thomas.koberda@yale.edu

J. Mangahas
Mathematics Department, Brown University, Providence, RI, USA
e-mail: mangahas@math.brown.edu

J. González-Meneses et al. (eds.), *Extended Abstracts Fall 2012*,
Trends in Mathematics 1, DOI 10.1007/978-3-319-05488-9_12,
© Springer International Publishing Switzerland 2014

on $\pi_1(S)$, or set of curves cutting S into disks and peripheral annuli) whose outputs differ only by quasi-isometry. The key to our algorithm is the following:

Theorem 1. *One can compute C depending on X and ℓ such that, for any mapping class $f \in \text{Mod}(S)$ and component α of $\sigma(f)$,*

$$\ell(\alpha) \leq C^{|f|}.$$

Existence of the exponential function above follows from the proof of solvability of the conjugacy problem for $\text{Mod}(S)$ (Theorem E in [10]), but writing it explicitly that way requires knowledge of constants related to the curve complex. On the other hand, when $X = \{T_{\gamma_i}\}$ consists of Dehn–Humphries–Lickorish generators, and curve length is defined by $\ell(\alpha) = \max_i i(\alpha, \gamma_i)$, our computation of C above gives

$$\ell(\alpha) \leq 2^{25\chi(S)|f|}. \tag{1}$$

Dehn–Humphries–Lickorish generators are optimal in that these are twists about curves that, pairwise, intersect at most once. Generally, one can finitely generate $\text{Mod}(S)$ with Dehn twist generators for $\text{PMod}(S)$ (the subgroup of mapping classes that do not permute punctures) plus *half twists* about separating curves; in that case one can arrange for (1) except with the base of the exponent being 3 instead of 2. Since our methods do not recognize boundary twists, all boundary components must be treated as punctures.

Given Theorem 1, the algorithm proceeds as follows. We are given a word f in the generators X. Let $F = C^{|f|}$ be the upper bound on the length of components of $\sigma(f)$. Let N be a power, depending only on S, that ensures f^N fixes individually, rather than permutes, the components of $\sigma(f)$; in particular, we may take $N = \xi(S)!$, where $\xi(S) = 3g - 3 + p$, the maximum number of disjoint curves on S. The steps are:

1. Populate the set LIST with the words representing all simple closed curves of length at most $4F$.
2. For each curve $\delta \in$ LIST, test if $f^N(\delta) = \delta$. If this is true, add δ to the set TEST-LIST.
3. Let SHORT-LIST be the subset of curves in TEST-LIST with length at most F.
4. For each curve $\delta \in$ SHORT-LIST, test for intersection with curves in TEST-LIST. If there are no intersections, then $\delta \in \sigma(f)$.
5. If $\sigma(f)$ is not empty, then f is reducible and its canonical reduction system is given. If $\sigma(f)$ is empty but SHORT-LIST is not empty, then f is finite-order. Otherwise f is pseudo-Anosov.

An alternate classification method uses Theorem 1 to make effective the result in [8] that, given $f \in \text{Mod}(S)$, one may find a finite cover $\tilde{S}$ and a lift $\tilde{f}$ of f to $\tilde{S}$, such that the action of $\tilde{f}_*$ on $H_1(\tilde{S}, \mathbb{C})$ determines the Nielsen–Thurston classification of f.

Sketch of proof of Theorem 1. For simplicity, let us assume S is closed and X consists of Dehn–Humphries–Lickorish generators $\{T_{\gamma_i}\}$. Define $\ell(\alpha) = \max_i i(\alpha, \gamma_i)$. One can easily show that

$$2^{|f|-1} \geq i(f(\gamma), \gamma) \tag{2}$$

for any $\gamma \in \{\gamma_i\}$, using an inductive argument and the well-known inequality about intersection number after Dehn twisting.

Now suppose α is a component of $\sigma(f)$. There are two possibilities: either α is the boundary of a subsurface on which some power f^p acts as a pseudo-Anosov, or α is the core curve around which some power f^p acts as a twist. The point is to show that, in either case, every arc of γ intersecting α guarantees some minimal number of intersections between $f^p(\gamma)$ and γ. For example, in the subsurface pseudo-Anosov case,

$$i(f^p(\gamma), \gamma) \geq [i(\alpha, \gamma)/(6|\chi(S')|)]^2. \tag{3}$$

With an efficient bound for the power p, one can derive an inequality like (1) using (2) and (3).   □

References

1. D. Benardete, M. Gutierrez, Z. Nitecki, A combinatorial approach to reducibility of mapping classes, in *Mapping Class Groups and Moduli Spaces of Riemann Surfaces* (Göttingen/Seattle, 1991). Contemporary Mathematics, vol. 150 (American Mathematical Society, Providence, 1993), pp. 1–31
2. D. Benardete, M. Gutierrez, Z. Nitecki, Braids and the Nielsen–Thurston classification. J. Knot Theory Ramif. **4**, 549–618 (1995)
3. M. Bestvina, M. Handel, Train-tracks for surface homeomorphisms. Topology **34**(1), 109–140 (1995)
4. M. Calvez, Fast Nielsen–Thurston classification of braids, arxiv:1112.0165v2[math.GT]
5. M. Calvez, B. Wiest, Fast algorithmic Nielsen–Thurston classification of four-strand braids. J. Knot Theory Ramif. **21**, 1250043 (2012) [25 pages] DOI:10.1142/S0218216511009959
6. J. González-Meneses, B. Wiest, Reducible braids and Garside theory. Algebr. Geom. Topol. **11**, 2971–3010 (2011)
7. H. Hamidi-Tehrani, Z. Chen, Surface diffeomorphisms via train-tracks. Topol. Appl. **73**, 141–167 (1996)
8. T. Koberda, Asymptotic linearity of the mapping class group and a homological version of the Nielsen–Thurston classification. Geom. Dedic. **156**, 13–30 (2012)
9. J. Los, Pseudo-Anosov maps and invariant train tracks in the disc: a finite algorithm. Proc. Lond. Math. Soc. **66**(2), 400–430 (1993)
10. J. Tao, Linearly bounded conjugator property for mapping class groups. Geom. Funct. Anal. **23**(1), 415–466 (2013)

Tree-Irreducible Automorphisms of Free Groups

Martin Lustig

Throughout this extended abstract of [7] we denote by F_N the non-abelian free group of finite rank $N \geq 2$. We assume familiarity of the reader with the basic terminology around automorphisms of free groups and Outer space; details as well as some back ground can be found for example in [6] or [8].

An outer automorphism $\varphi \in \mathrm{Out}(F_N)$ is called *iwip* (or *fully irreducible*) if no positive power φ^t fixes the conjugacy class of any non-trivial proper free factor $F_M \subset F_N$. Furthermore, φ is called *atoroidal* if no positive power φ^t fixes a non-trivial conjugacy class of elements of F_N. An automorphism of F_N is atoroidal if and only if it is *hyperbolic* (i.e., the associated mapping torus group $F_N \rtimes_\varphi \mathbb{Z}$ is Gromov-hyperbolic). An iwip φ is not atoroidal if and only if φ is *surface* (i.e., φ is induced by a homeomorphism of a surface with boundary).

It follows directly from these definitions that $\varphi \in \mathrm{Out}(F_N)$ is iwip if and only if φ^{-1} is iwip as well, and that φ is atoroidal if and only if φ^{-1} is also atoroidal.

Atoroidal iwip automorphisms have been widely recognized as the closest possible analogues in $\mathrm{Out}(F_N)$ of pseudo-Anosov mapping classes of surfaces. In particular, the following well known properties correspond directly to known properties of pseudo-Anosov mapping classes of a closed surface Σ, if one "translates" F_N into $\pi_1\Sigma$, "basis" into "standard generating system", "Outer space" into "Teichmüller space", etc.

Facts 1 ([1,3–5]). Let $\varphi \in \mathrm{Out}(F_N)$ be an iwip automorphism. Then the following holds:

(1) There exists a *stretching factor* $\lambda_\varphi > 1$ such that every conjugacy class $[w] \subset F_N$ which is not φ-periodic has uniform exponential growth

M. Lustig (✉)
LATP, Centre de Mathématiques et Informatique, Aix-Marseille Université, Marseille, France
e-mail: Martin.Lustig@univ-amu.fr

J. González-Meneses et al. (eds.), *Extended Abstracts Fall 2012*,
Trends in Mathematics 1, DOI 10.1007/978-3-319-05488-9_13,
© Springer International Publishing Switzerland 2014

$$\widehat{|\varphi^t([w])|}_{\mathcal{A}} \sim \lambda_\varphi^t C_{[w]}^{\mathcal{A}}.$$

Here $\mathcal{A}$ is any "fixed" basis of F_N, $C_{[w]}^{\mathcal{A}} > 0$ is a constant which depends only on $[w]$ (and the choice of $\mathcal{A}$), $|\cdot|_{\mathcal{A}}$ denotes word length in $\mathcal{A} \cup \mathcal{A}^{-1}$, and by $\hat{v}$ we mean the cyclically reduced word in $\mathcal{A}$ which is given (up to cyclic permutation) by the conjugacy class $[v] \in F_N$.

(Moreover, if φ is atoroidal then the only φ-periodic conjugacy class is the trivial one.)

(2) φ is represented by an (absolute) train track map $f : \Gamma \to \Gamma$, with respect to some marking isomorphism $\pi_1 \Gamma \cong F_N$. The non-negative transition matrix $M(f)$ (defined by the action of f on the edges of Γ) is primitive, and its Perron–Frobenius eigenvalue is equal to λ_φ.

(3) φ does not split over any very small graph-of-groups decomposition of F_N. Equivalently, there is no very small simplicial tree T with F_N-action by homeomorphisms such that φ "commutes" with certain homeomorphisms $H : T \to T$. (By this we mean that there exists a representative $\Phi \in \mathrm{Aut}(F_N)$ of φ which satisfies $\Phi(w)H = Hw : T \to T$ for every $w \in F_N$.)

(4) There is an (up to rescaling) unique $\mathbb{R}$-tree $T_+ = T_+(\varphi)$, equipped with a very small action of F_N by isometries, which is projectively φ-invariant with stretching factor $\lambda_+ > 1$:

$$||\varphi([w])||_{T_+} = \lambda_+ ||[w]||_{T_+}$$

for any $w \in F_N$, where $||\cdot||_{T_+}$ denotes the translation length on T_+. Furthermore one has $\lambda_+ = \lambda_\varphi$ (as given in (1) above).

(5) The tree $T_+(\varphi)$ is indecomposable. This means that for any two non-degenerate segments $I, J \subset T$ the segment J is contained in the union of finitely many translates $w_i \overset{\circ}{I}$ of the interior $\overset{\circ}{I}$ of I, with $w_i \overset{\circ}{I} \cap w_{i+1} \overset{\circ}{I} \neq \emptyset$.

(Moreover, if φ is atoroidal, then the action of F_N on $T_+(\varphi)$ is free.)

(6) φ induces on the Thurston compactification $\overline{CV}_N$ of Culler–Vogtmann's outer space CV_N a locally uniform North-South dynamics, where the sink is the projective class $[T_+(\varphi)]$, and the source is the projective class $[T_-(\varphi)]$, with $T_-(\varphi) = T_+(\varphi^{-1})$.

Since, as pointed out above, the assumption "φ (atoroidal) iwip" is equivalent to "φ^{-1} (atoroidal) iwip", all of the above facts are true for φ^{-1} as well. The reader should however be warned that, contrary to the situation of pseudo-Anosov's on surfaces, the stretching factor $\lambda_- := \lambda_{\varphi^{-1}}$ may well be different from $\lambda_+ = \lambda_\varphi$. For more information about the fine-structure of iwips and the subtle differences to pseudo-Anosov homeomorphisms, see [2].

If one considers a surface Σ with precisely one boundary component, and a pseudo-Anosov homeomorphism $h : \Sigma \to \Sigma$, then the induced outer automorphism $\varphi = h_*$ on $F_N = \pi_1 \Sigma$ is iwip, but not atoroidal. Indeed, all iwips φ which are not atoroidal arise precisely in this way.

Facts 2. Let $\varphi \in \mathrm{Out}(F_N)$ be induced by a pseudo-Anosov homeomorphism $h: \Sigma \to \Sigma$ of a surface Σ with 1 or more boundary components, via some isomorphism $F_N \cong \pi_1 \Sigma$. Then one has:

(1) φ satisfies all the properties (1)–(6) of Facts 1.
(2) However, if Σ has at least 2 boundary components, then φ is not fully irreducible ($=$ iwip).

The proof of fact (2) is obvious, since each boundary curve C_i of Σ generates a cyclic subgroup which, in the presence of at least one more boundary component $C_j \neq C_i$, is a free factor of $\pi_1 \Sigma$, and some power of h maps C_i to itself. The reader should note here that, contrary to the situation for pseudo-Anosov homeomorphisms on a surface, the negation of "iwip" is for a free group automorphisms φ in general a much weaker property than the existence of a φ-invariant small graph-of-groups decomposition of F_N.

In view of these facts it seems clear that for many purposes the notion of iwip automorphisms is too restrictive. We propose here the following definition as a useful class of free group automorphisms:

Definition 3. An automorphism $\psi \in \mathrm{Out}(F_N)$ is called *tree-irreducible* if the following properties are satisfied:

(1) There is no very small graph-of-groups decomposition of F_N which is preserved by ψ.
(2) There is an (up to rescaling) unique $\mathbb{R}$-tree $T_+ = T_+(\psi)$ with very small action of F_N by isometries that is projectively ψ-invariant with stretching factor $\lambda_+ > 1$.
(3) The tree $T_+(\psi)$ is indecomposable.
    (However, the action of F_N on $T_+(\psi)$ is in general not free.)
(4) The outer automorphism ψ fixes every conjugacy class of any elliptic subgroup of T. Equivalently, if $x \in T$ is a point with non-trivial stabilizer $\mathrm{stab}_{F_N}(x) \subset F_N$, then φ fixes the F_N-orbit of the subgroup $\mathrm{stab}_{F_N}(x)$, and it induces on it the trivial outer automorphism.

It turns out that this definition is not minimal: one can deduce property (1) as well as the uniqueness in property (2) from the remaining properties in Definition 3. On the other hand, the following is not totally obvious:

Lemma 4. *The inverse of any tree-irreducible $\psi \in \mathrm{Out}(F_N)$ is also tree-irreducible.*

The properties (1) and (2) (the latter slightly adjusted) of Facts 1 also hold for tree-irreducible automorphisms. The most interesting property however, given its wide use for iwips, seems to be property (6). It turns out that property (3) of Definition 3 is the main ingredient in the proof given in [5] for iwips, so that its main part can be applied word-by-word to tree-irreducible automorphisms as well. We obtain:

Theorem 5. *Every tree-irreducible automorphism* $\psi \in \mathrm{Out}(F_N)$ *induces on the compactified Outer space* $\overline{CV}_N$ *a locally uniform North-South dynamics.*

As in the iwip case, the sink of this North-South dynamics is the projective class $[T_+(\psi)]$, and the source is the projective class $[T_-(\psi)]$, with $T_-(\psi) = T_+(\psi^{-1})$.

The second main reason why one may consider seriously tree-irreducible automorphisms of F_N comes from train-track technology: Bestvina–Handel, in their celebrated paper [1], have shown that every automorphism $\varphi \in \mathrm{Out}(F_N)$ possesses a relative train-track representative, i.e., there is a graph Γ with marking isomorphisms $F_N \cong \pi_1 \Gamma$ such that φ is induced by a self-map $f : \Gamma \to \Gamma$, and there exists an f-invariant filtration into non-necessarily connected subgraphs $\Gamma = \Gamma_r \supset \Gamma_{r-1} \supset \cdots \supset \Gamma_1 \supset \Gamma_0$, such that on each *stratum* Γ_i, considered modulo Γ_{i-1}, the restriction of f is either zero, periodic, or train track with primitive transition matrix. After rising f to a suitable positive power f^t it follows that φ^t can be written as *semi-commuting product* (see [7]) $\varphi_s \circ \varphi_{s-1} \circ \cdots \circ \varphi_1$, where each φ_i is either a Dehn twist automorphism, or else it is, up to an enlargement of the elliptic subgroups, a tree-irreducible automorphism.

The "enlargement technique" in the previous paragraph, however, is more delicate than one may suspect at first sight. In particular, we use this technique to exhibit the first explicit examples of the following kind:

Proposition 6. *There exist* $\mathbb{R}$-*trees* T *with very small action of* F_N *by isometries which are not indecomposable, but do also not decompose as dusted action or as very small graph-of-actions.*

The interest in this subtle phenomenon comes from the fact that the opposite statement, but with the words "very small" before "graph-of-actions" erased, has been recently proved by Guirardel–Levitt.

Finally, in the last part of [7] a technique is presented for producing tree-irreducible automorphisms from fully irreducible ones by "puncturing the singularities of the expanding lamination", in analogy to what is often done for pseudo-Anosov homeomorphisms of surfaces.

References

1. M. Bestvina, M. Handel, Train tracks and automorphisms of free groups. Ann. Math. **135**(2), 1–51 (1992)
2. T. Coulbois, A. Hilion, Botany of irreducible automorphisms of free groups. Pac. J. Math. **256**, 291–307 (2012)
3. T. Coulbois, A. Hilion, P. Reynolds, Indecomposable F_N-trees and minimal laminations, arXiv: math.GR1110.3506
4. I. Kapovich, M. Lustig, Invariant laminations for irreducible automorphisms of free groups. Quart. J. Math. first published online January 30, 2014 doi:10.1093/qmath/hat056 (35 pages)
5. G. Levitt, M. Lustig, Irreducible automorphisms of F_n have North-South dynamics on compactified outer space. J. Inst. Math. Jussieu **2**, 59–72 (2003)

6. M. Lustig, Conjugacy and centralizers for iwip automorphisms of free groups, in *Geometric Group Theory: Geneva and Barcelona Conferences*, ed. by G. Arzhantseva, L. Bartholdi, J. Burillo, E. Ventura. Trends in Mathematics. (Birkhäuser, Basel, 2007), pp. 197–224
7. M. Lustig, Tree-irreducible automorphisms of free groups preprint, (2013), available at http://arxiv.org/pdf/1306.5688.pdf
8. K. Vogtmann, Automorphisms of free groups and outer space. Geom. Dedic. **94**, 1–31 (2002)

Presentations for the Mapping Class Groups of Nonorientable Surfaces

Luis Paris

Let N be a nonorientable surface (possibly with boundary). We denote by $\mathcal{H}(N)$ the group of homeomorphisms $h\colon N \to N$ that pointwise fix the boundary of N. The *mapping class group* of N, denoted by $\mathcal{M}(N)$, is the group of isotopy classes of elements of $\mathcal{H}(N)$.

The aim of this talk is to give explicit presentations for the mapping class groups of nonorientable surfaces of genus $g \geq 3$ with one boundary component. It is based on a joint work with Blazej Szepietowski [4].

A *simple closed curve* is an embedding $\gamma\colon \mathbb{S}^1 \to N$. We assume that γ is oriented, and we denote by γ^{-1} the curve having the same image but opposite orientation. We say that γ is *generic* if it does not bound neither a disc, nor a Moebius band (cross-cap) in N. We say that γ is *two-sided* if the regular neighborhood of γ is an annulus, and we call it *one-sided* if its regular neighborhood is a Moebius band. We say that γ is *nonseparating* if $N \setminus \gamma$ is connected.

Let $C = (\gamma_1, \ldots, \gamma_m)$ be an m-tuple of generic curves. We say that C is a *generic m-tuple of curves* if, for $i \neq j$,

(a) γ_i is disjoint from γ_j;
(b) γ_i is not isotopic to $\gamma_j^{\pm 1}$.

Two generic m-tuples $C = (\gamma_1, \ldots, \gamma_m)$ and $C' = (\gamma_1', \ldots, \gamma_m')$ are *equivalent* if γ_i is isotopic to $\gamma_i'^{\pm 1}$ for all $i \in \{1, \ldots, m\}$. Note that the ordering of the curves in C is important for us. If $C = (\gamma_1, \ldots, \gamma_m)$ is a generic m-tuple of curves, we denote by N_C the surface obtained by cutting N along the curves $\gamma_1, \ldots, \gamma_m$.

The *ordered complex of curves* of N is the Δ-complex $\mathcal{C}(N)$ such that

(a) An m-simplex is an equivalence class of generic $(m+1)$-tuples of curves, $C = (\gamma_0, \ldots, \gamma_m)$;

L. Paris (✉)
Institut de Mathématiques de Bourgogne, Université de Bourgogne, Dijon, France
e-mail: lparis@u-bourgogne.fr

J. González-Meneses et al. (eds.), *Extended Abstracts Fall 2012*,
Trends in Mathematics 1, DOI 10.1007/978-3-319-05488-9_14,
© Springer International Publishing Switzerland 2014

Fig. 1 Representation of a
nonorientable surface

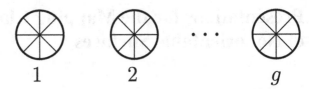

$$1 \qquad\qquad 2 \qquad\qquad\qquad g$$

(b) The faces of C are the $(m-1)$-simplices $(\gamma_0,\ldots,\hat{\gamma}_i,\ldots,\gamma_m)$, $i = 0,\ldots,m$.

We denote by $\mathcal{C}_0(N)$ the subcomplex of $\mathcal{C}(N)$ made of the classes C such that N_C is connected. We denote by $\mathcal{D}(N)$ the subcomplex of $\mathcal{C}_0(N)$ made of the classes C such that N_C is nonorientable.

Combining results from [6] with techniques from [1], we get the following.

Theorem 1 (Paris–Szepietowski [4]).

(1) $\mathcal{C}_0(N_{g,n})$ *is simply connected if* $g \geq 5$.
(2) $\mathcal{D}(N_{g,n})$ *is simply connected if* $g \geq 7$.

By applying the following theorem to $\mathcal{C}_0(N_{g,1})$ if $g = 5, 6$, and to $\mathcal{D}(N_{g,1})$ if $g \geq 7$, we calculate an explicit presentation for $\mathcal{M}(N_{g,1})$ using induction on g (plus many calculations). Presentations for $\mathcal{M}(N_{g,1})$ for $g \leq 4$ can be found in [5].

Theorem 2 (Brown [2]). *Let G be a group acting on a simply connected CW-complex X. We assume that the action of G on X sends m-cells to m-cells. We assume also that:*

(a) *The isotropy subgroup of each vertex if finitely presented;*
(b) *The isotropy subgroup of each edge is finitely generated;*
(c) *G has finitely many orbits of 2-cells.*

Then G is finitely presented and there is a method for calculating a presentation for G from this data.

We turn now to describe the presentation for $\mathcal{M}(N_{g,1})$.

We picture $N = N_{g,1}$ as follows (see Fig. 1). We remove the interior of g closed discs to a given disc. We get in this way an oriented surface of genus 0 with $g + 1$ boundary components. Then we glue cross-caps to this new g boundary components.

Let $a_1,\ldots,a_{g-1}, b$ be the curves represented in Fig. 2. Observe that each a_i is a two-sided curve. The curve b is also two-sided. We denote by A_i the Dehn twist along a_i, and by B the Dehn twists along b. Let S be a regular neighborhood of $a_1 \cup \cdots \cup a_{g-1}$. Then S is an oriented surface of genus $\frac{1}{2}(g - 1)$ with one boundary component if g is odd, and it is an oriented surface of genus $\frac{1}{2}(g - 2)$ with two boundary components if g is even. The embedding $S \hookrightarrow N$ induces an injective homomorphisms $\mathcal{M}(S) \to \mathcal{M}(N)$. Moreover, $A_1,\ldots,A_{g-1}, B$ generates $\mathcal{M}(S)$ and a presentation for $\mathcal{M}(S)$ with these generators can be easily calculated from the presentation given in [3].

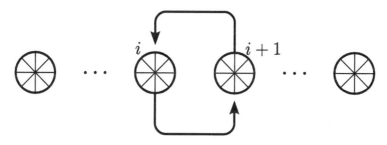

Fig. 2 Curves in N

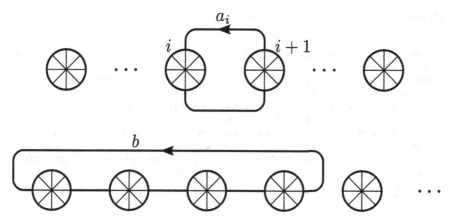

Fig. 3 The transformation U_i

We set an orientation on a_i as in Fig. 2. We denote by U_i the transformation which transports the i-th cross-cap on the $(i + 1)$-th cross-cap along a_i, and which transports the $(i + 1)$-th cross-cap on the i-th cross-cap along a_i (see Fig. 3). Note that U_i reverses the orientation in a regular neighborhood of a_i. Observe also that $U_1, \ldots, U_{g-1}$ generate a subgroup of $\mathcal{M}(N)$ isomorphic to $\mathcal{B}_g$, the braid group on g strands.

Theorem 3 (Paris–Szepietowski [4]). *The group* $\mathcal{M}(N)$ *is isomorphic to the quotient of* $\mathcal{M}(S) * \mathcal{B}_g$ *by the following relations:*

(C1) $A_i U_j = U_j A_i$, *for* $|i - j| \geq 2$,
(C2) $A_i U_{i+1} U_i = U_{i+1} U_i A_{i+1}$, *for* $1 \leq i \leq g - 2$,
(C3) $A_{i+1} U_i U_{i+1} = U_i U_{i+1} A_i$, *for* $1 \leq i \leq g - 2$,
(C4) $A_i U_i A_i = U_i$, *for* $1 \leq i \leq g - 1$,
(C5) $U_{i+1} A_i A_{i+1} U_i = A_i A_{i+1}$, *for* $1 \leq i \leq g - 2$,
(C6) $(U_3 B)^2 = (A_1 A_2 A_3)^2 (U_1 U_2 U_3)^2$,
(C7) $U_5 B = B U_5$,
(C8) $A_4 U_4 A_4 A_3 A_2 A_1 U_1 U_2 U_3 U_4 B = B A_4 U_4$.

References

1. S. Benvenuti, Finite presentations for the mapping class group via the ordered complex of curves. Adv. Geom. **1**(3), 291–321 (2001)
2. K.S. Brown, Presentations for groups acting on simply-connected complexes. J. Pure Appl. Algebra **32**(1), 1–10 (1984)
3. C. Labruere, L. Paris, Presentations for the punctured mapping class groups in terms of Artin groups. Algebr. Geom. Topol. **1**, 73–114 (2001)
4. L. Paris, B. Szepietowski, A presentation for the mapping class group of a nonorientable surface, preprint, available at http://arxiv.org/pdf/1308.5856.pdf
5. B. Szepietowski, A presentation for the mapping class group of the closed non-orientable surface of genus 4. J. Pure Appl. Algebra **213**(11), 2001–2016 (2009)
6. N. Wahl, Homological stability for the mapping class groups of non-orientable surfaces. Invent. Math. **171**(2), 389–424 (2008)

Ideal Whitehead Graphs in Out(F_r)

Catherine Pfaff

For a compact surface S, the *mapping class group* $\mathcal{MCG}(S)$ is the group of isotopy classes of homeomorphisms $h: S \rightarrow S$. A generic (see, for example, [7]) mapping class is *pseudo-Anosov*, i.e., has a representative leaving invariant a pair of transverse measured singular minimal foliations. From the foliation comes a singularity index list. Masur and Smillie determined precisely which of the singularity index lists, permitted by the Poincaré–Hopf index formula, arise from pseudo-Anosovs [8]. The index lists were significant in their stratification of the space of quadratic differentials, with strata invariant under the Teichmüller flow.

For a free group F_r of rank r, we denote the outer automorphism group by Out(F_r). In this paper, as in [2] for example, we analyze outer automorphisms by topological representatives: Let R_r be the r-petaled rose. Given a connected graph Γ with $\pi(\Gamma) \cong F_r$ and no valence-one vertices, we can assign to Γ a *marking* via a homotopy equivalence $R_r \rightarrow \Gamma$. One calls such a graph, together with its marking, a *marked graph* [2]. Each outer automorphism $\phi \in \text{Out}(F_r)$ can be represented by a homotopy equivalence $g: \Gamma \rightarrow \Gamma$ of a marked graph, where $\phi = g_*$ is the induced map of fundamental groups. Analogous to pseudo-Anosov mapping classes are *fully irreducible (iwip)* outer automorphisms, i.e., those such that no representative of any power leaves invariant a subgraph with a nontrivial component. Thus, an analog to the [8] theorem would involve fully irreducible outer automorphisms.

A beauty in studying the groups Out(F_r) is how they are actually richly more complicated than mapping class groups. A particularly good example of this arises when trying to generalize the Masur–Smillie pseudo-Anosov index theorem. Unlike in the surface case where one has the Poincaré–Hopf index equality $i(\psi) = \chi(S)$, for a pseudo-Anosov ψ on a surface S, Gaboriau, Jaeger, Levitt, and Lustig proved [4] that there is instead an index sum inequality $0 \geq i(\phi) \geq 1 - r$ for the fully

C. Pfaff (✉)
LATP, Centre de Mathématiques et Informatique, Aix-Marseille Université, Marseille, France
e-mail: catherine.pfaff@latp.univ-mrs.fr

J. González-Meneses et al. (eds.), *Extended Abstracts Fall 2012*,
Trends in Mathematics 1, DOI 10.1007/978-3-319-05488-9_15,
© Springer International Publishing Switzerland 2014

irreducible $\phi \in \text{Out}(F_r)$. Here we have switched the sign in the [4] index definition (thus inequality) for consistency with the surface case.

The index lists of *geometric* fully irreducibles (those induced by homeomorphisms of compact surfaces with boundary) are understood by the [8] theorem, but complexity of the nongeometric case prompted the following question by Handel and Mosher [5]:

Question 1. Which index types, satisfying $0 \geq i(\phi) > 1 - r$, are achieved by nongeometric, fully irreducible $\phi \in \text{Out}(F_r)$?

It is proved in [3] that for any fully irreducible there exists a *rotationless* power fixing all periodic directions. For this reason, and because we focus on lamination invariants, instead of precisely using the [4] index definition, we use a definition reliant on periodic points at infinity instead of just fixed points at infinity. Still we make indices negative. We call this the *rotationless index*. Note that it is invariant under taking powers and that, when restricting to rotationless powers, the two definitions are the same up to sign. While we do not directly use the index definition here, it is implicit in our notation.

By [4] we know there are two kinds of nongeometric fully irreducibles, namely "ageometrics" and "parageometrics". Also [4], with the rotationless index, equality holds for parageometrics, leaving us to analyze ageometrics: Recall [1], for a train track map $g: \Gamma \to \Gamma$, a *periodic Nielsen path (pNp)* is a nontrivial path ρ in Γ such that, for some k, $g^k(\rho) \simeq \rho$ rel endpoints. By [4], an outer automorphism is *ageometric* whose stable representative, in the sense of [2], has no pNp's (closed or otherwise).

Beyond the existence of an inequality, instead of just an equality, "ideal Whitehead graphs" give yet another layer of complexity for fully irreducibles. In the surface case, ideal Whitehead graphs are all circles. However, as mentioned before, the ideal Whitehead graph $\mathcal{IW}(\phi)$ for a fully irreducible $\phi \in \text{Out}(F_r)$ (see [5]) gives a strictly finer outer automorphism invariant than just the corresponding index list. Indeed, each connected component C_i of $\mathcal{IW}(\phi)$ contributes the index $1 - \frac{1}{2}k_i$ to the list, where C_i has k_i vertices. One can see many complicated ideal Whitehead graph examples. For example, we prove in [9] the existence of the complete graph in each rank (see Theorem 6 below). Additionally, we shown in [10] that 18 of the 21 connected, simplicial 5-vertex graphs are achieved by fully irreducibles in rank-3. The deeper, more appropriate question is thus:

Question 2. Which isomorphism types of graphs occur as the ideal Whitehead graph $\mathcal{IW}(\phi)$ of a fully irreducible outer automorphism ϕ?

Pfaff [10] gives a complete answer to Question 2 in rank three for the single-element index list $(-\frac{3}{2})$ (see Theorem 4 below for the graphs that are not achieved):

Theorem 3. *Precisely 18 of the 21 connected, simplicial 5-vertex graphs are the ideal Whitehead graph $\mathcal{IW}(\phi)$ for a fully irreducible outer automorphism $\phi \in \text{Out}(F_3)$.*

However, our first step (given in [11]) was a theorem providing examples in each rank of connected $(2r - 1)$-vertex graphs that are not the ideal Whitehead graph $\mathcal{IW}(\phi)$ for any fully irreducible $\phi \in \text{Out}(F_r)$, i.e., that are *unachieved* in rank r:

Theorem 4. *For each $r \geq 3$, let $\mathcal{G}_r$ be the graph consisting of $2r - 2$ edges adjoined at a single vertex.*

(a) *For no fully irreducible $\phi \in \text{Out}(F_r)$ is $\mathcal{IW}(\phi) \cong \mathcal{G}_r$.*

(b) *The following connected graphs are not the ideal Whitehead graph $\mathcal{IW}(\phi)$ for any fully irreducible $\phi \in \text{Out}(F_3)$:*

Our proof of this theorem required our introduction of a "Birecurrency Condition" (see [11]), a proof of the existence of "ideal decompositions", and our development of automata, called ideal "decomposition ($\mathcal{ID}$) diagrams". A key fact we prove is that every ideally decomposed representative is realized by a loop in an $\mathcal{ID}$ diagram.

We state here the ideal decomposition existence proposition without defining all relevant terms. What is says is that, given an ageometric, fully irreducible $\phi \in \text{Out}(F_r)$ whose ideal Whitehead graph $\mathcal{IW}(\phi)$ is a connected, $(2r - 1)$-vertex simplicial graph, some power ϕ^R has a rotationless train track representative whose Stallings fold decomposition (see [12]) consists entirely of proper full folds of roses.

Proposition 5. *Let $\phi \in \text{Out}(F_r)$ be an ageometric, fully irreducible outer automorphism whose ideal Whitehead graph $\mathcal{IW}(\phi)$ is a connected, $(2r - 1)$-vertex graph. Then there exists a train track representative of a power $\psi = \phi^R$ of ϕ that*

(1) *Is on the rose,*

(2) *Is rotationless,*

(3) *Has no periodic Nielsen paths, and*

(4) *Is decomposable as a sequence of proper full folds of roses.*

In fact, it decomposes as $\Gamma = \Gamma_0 \xrightarrow{g_1} \Gamma_1 \xrightarrow{g_2} \cdots \xrightarrow{g_{n-1}} \Gamma_{n-1} \xrightarrow{g_n} \Gamma_n = \Gamma$, where

(I) *The index set $\{1, \ldots, n\}$ is viewed as the set $\mathbf{Z}/n\mathbf{Z}$ with its natural cyclic ordering;*

(II) *Each Γ_k is an edge-indexed rose with an indexing*

$$\{e_{(k,1)}, e_{(k,2)}, \ldots, e_{(k,2r-1)}, e_{(k,2r)}\},$$

where

(a) *One can edge-index Γ with $\mathcal{E}(\Gamma) = \{e_1, e_2, \ldots, e_{2r-1}, e_{2r}\}$ such that, for each t with $1 \leq t \leq 2r$, $g(e_t) = e_{i_1} \ldots e_{i_s}$ where $(g_n \circ \cdots \circ g_1)(e_{0,t}) = e_{n,i_1} \ldots e_{n,i_s}$;*

(b) *For some i_k, j_k with $e_{k,i_k} \neq (e_{k,j_k})^{\pm 1}$*

$$g_k(e_{k-1,t}) := \begin{cases} e_{k,t} e_{k,j_k} \text{ for } t = i_k \\ e_{k,t} \text{ for all } e_{k-1,t} \neq e_{k-1,j_k}^{\pm 1}; \text{ and} \end{cases}$$

(the edge index permutation for the homeomorphism in the decomposition is trivial, so it is left out)

(c) *For each $e_t \in \mathcal{E}(\Gamma)$ with $t \neq j_n$, we have $Dh(d_t) = d_t$, where $d_t = D_0(e_t)$.*

The representatives of Proposition 5 are called "ideally decomposable". Our second step was the existence of the complete graph in each rank [9]:

Theorem 6. *For each $r \geq 3$, let C_r denote the complete $(2r-1)$-vertex graph. Then, for each $r \geq 3$, there exists an ageometric, fully irreducible outer automorphism $\phi \in \text{Out}(F_r)$ such that C_r is the ideal Whitehead graph $\mathcal{IW}(\phi)$ for ϕ.*

In order to show that our examples for Theorem 6 indeed represented fully irreducible outer automorphisms, we proved in [10] a folk lemma, Lemma 7, the "Full Irreducibility Criterion (FIC)". In [6], another criterion is given inspired by our FIC.

Lemma 7 (The Full Irreducibility Criterion). *Let g be a train track representing an outer automorphism $\phi \in \text{Out}(F_r)$ such that*

(I) *g has no periodic Nielsen paths,*

(II) *The transition matrix for g is Perron–Frobenius, and*

(III) *All local Whitehead graphs $\mathcal{LW}(x; g)$ for g are connected.*

Then ϕ is fully irreducible.

In order to apply the criterion, we gave in [10] a method for identifying periodic Nielsen paths.

For a fully irreducible $\phi \in \text{Out}(F_r)$, to have the index list $(\frac{3}{2} - r)$, ϕ must be ageometric with a connected, $(2r - 1)$-vertex ideal Whitehead graph $\mathcal{IW}(\phi)$. We chose to focus on the single-element index list $(\frac{3}{2} - r)$ because it is the closest to that achieved by geometrics, without being achieved by a geometric.

References

1. M. Bestvina, M. Feighn, Outer limits (1994, preprint). Available at http://andromeda.rutgers.edu/~feighn/papers/outer.pdf
2. M. Bestvina, M. Handel, Train tracks and automorphisms of free groups. Ann. Math. **135**(1), 1–51 (1992)

3. M. Feighn, M. Handel, The recognition theorem for Out(F_n). Groups Geom. Dyn. **5**(1), 39–106 (2011)

4. D. Gaboriau, A. Jaeger, G. Levitt, M. Lustig, An index for counting fixed points of automorphisms of free groups. Duke Math. J. **93**(3), 425–452 (1998)

5. M. Handel, L. Mosher, *Axes in Outer Space*. Memoirs of the American Mathematical Society, vol. 123, no. 1004 (American Mathematical Society, Providence, 2011)

6. I. Kapovich, Algorithmic detectability of iwip automorphisms (2012), arXiv:1209.3732

7. J. Maher, Random walks on the mapping class group. Duke Math. J. **156**(3), 429–468 (2011)

8. H. Masur, J. Smillie, Quadratic differentials with prescribed singularities and pseudo-Anosov diffeomorphisms. Comment. Math. Helv. **68**(1), 289–307 (1993)

9. C. Pfaff, Ideal whitehead graphs in Out(F_r) II: complete graphs in every rank. Journal of Homotopy and Related Structures, (2013) pp. 1–27. Available at http://dx.doi.org/10.1007/s40062-013-0060-5

10. C. Pfaff, Ideal whitehead graphs in Out(F_r) III: achieved graphs in rank 3, preprint, available at http://arxiv.org/pdf/1301.7080v1.pdf

11. C. Pfaff, Ideal whitehead graphs in Out(F_r) I: some unachieved graphs, preprint available at http://arxiv.org/pdf/1210.5762v1.pdf

12. J.R. Stallings, Topology of finite graphs. Invent. Math. **71**(3), 551–565 (1983)

Reflections on Computer Experiments with Automorphisms of Free Groups

Paul Schupp

(...) I have thought fit to write out for you and explain in detail (...) a certain method, with which furnished, you will be able to make a beginning in the investigation by *mechanics* of some of the problems in mathematics. (...) This method is no less useful even for the proofs of the theorems themselves. For some things first became clear to me by *mechanics*, though they had to be proved geometrically owing to the fact that investigation by this method does not amount to actual proof; but it is, of course, easier to provide the proof when some knowledge of the things sought has been acquired by this method rather than to seek it with no prior knowledge.

Archimedes, *The Method*, in: *Greek Mathematical Works II. From Aristarchus to Pappus*, Loeb Classical Library, 362, Harvard University Press, 1993, 221–223. Quoted in Polkinghorne (ed.), *Meaning in Mathematics*.

If one replaces the word *mechanics* in the quote from Archimedes by the words *computer experiments* it then exactly states the philosophy of this talk. I will review some results about automorphisms of free groups obtained in joint work with Ilya Kapovich and other colleagues. All of these results were motivated by computer experiments. The topics considered will be the following:

1. Asymptotic density in free groups.
2. Test elements in the free group of rank 2.
3. Genericity and strict minimality under automorphism.
4. Rewriting 2-generator one-relator groups and mapping tori of automorphisms.
5. Translation equivalence and trace identities
6. Stretching factors for free group automorphisms.

P. Schupp (✉)

Department of Mathematics, University of Illinois at Urbana-Champaign, Urbana, IL, USA

e-mail: schupp@math.uiuc.edu

J. González-Meneses et al. (eds.), *Extended Abstracts Fall 2012*,

Trends in Mathematics 1, DOI 10.1007/978-3-319-05488-9_16,

© Springer International Publishing Switzerland 2014

1 Asymptotic Density of Subsets of Free Groups

If $F = \langle x_1, \ldots, x_n \rangle$ is a free group of rank n, then the *group alphabet* is

$$\Sigma = \langle x_1, \ldots, x_n, x_1^{-1}, \ldots, x_n^{-1} \rangle.$$

Every element of F has a unique representative as a freely reduced word. We measure the asymptotic density of subsets of F in the following way.

Definition 1. If $T \subseteq F$, then

$$\rho_n(T) = \frac{|\{w \in T, |w| \le n\}|}{|\{w \in F, |w| \le n\}|}.$$

If $\rho(T) = \lim_{n \to \infty} \rho_n(T)$ exists, then this limit is called the *asymptotic density* of T in F.

Definition 2. T is *generic* if $\rho(T) = 1$, and T is *negligible* if $\rho(T) = 0$.

2 Annular Density

For the next section we will also need the idea of annular density.

Definition 3. The *sphere* $S(n)$ of radius n in the free group is $\{w : |w| = n\}$. If $T \subseteq F$, define

$$\gamma_n(T) = \frac{1}{2} \frac{|\{w : w \in T, |w| = n - 1\}| + |\{w : w \in T, |w| = n\}|}{|\{w : |w| = n - 1\}| + |\{w : |w| = n\}|}.$$

If $\gamma(T) = \lim_{n \to \infty} \gamma_n(T)$ exists, then this limit is called the *annular density* of T in F.

Example 4. It is clear that if we take T to be the set of words of even length, then the asymptotic density of T does not exist, but the annular density of T is $\frac{1}{2}$.

3 Test Elements in the Free Group of Rank 2

Definition 5. Let $\mathcal{A}$ be an algebra, in the sense of universal algebra. An element $w \in \mathcal{A}$ is called a *test element* if every endomorphism of $\mathcal{A}$ which fixes w is must actually be an automorphism of $\mathcal{A}$.

The following theorem is proved in [5]:

Theorem 6. *In F_2, the free group of rank 2, the annular density of test elements is equal to $1 - 6/\pi^2$.*

It was clear by computer experiment that this was the *correct* answer. An experiment with a larger sample and longer words picked up another digit or two of the decimal expansion.

Definition 7. Let F_k be the free group on $\{x_1, \ldots, x_k\}$ and let $\alpha\colon F \to \mathbb{Z}_k$ be the abelianization map $\alpha(x_i) = e_i$. A nonzero element z of $\mathbb{Z}_k$ is called *visible* if the greatest common divisor of the coordinates of z is 1. (This terminology reflects the fact that if z is visible then the line segment from 0 to z does not contain any other integer lattice points.) We "lift" this terminology to F_k by defining an element w of F_k to be *visible* if $\alpha(w)$ is visible in $\mathbb{Z}_k$.

Fact 8. The probability that a random pair of integers is relatively prime is

$$\frac{1}{\zeta(2)} = \frac{6}{\pi^2}.$$

So the probability that a random element of $\mathbb{Z}_2$ is visible is $6/\pi^2$.

Proposition 9. *Let F_2 be the free group of rank 2. An element $w \in F_2$ which is not a proper power is not a test element if and only if w is visible in F_2.*

The technical work of the paper is to prove that one can lift asymptotic densities of suitable sets in $\mathbb{Z}_2$ back to the annular density of their preimage in F_k.

4 Genericity and Strict Minimality Under Automorphism

The *automorphism problem* for the free group F_k of rank k is the following decision problem. Is there an algorithm which, when given an arbitrary pair (u, v) of elements of F_k, decides whether or not there exists some automorphism τ of F_k such that $\tau(u) = v$?

J. H. C. Whitehead gave an algorithm solving this problem in 1936. We need an outline of Whitehead's algorithm.

A *Whitehead automorphism* of F_k is an automorphism τ of one of the following two kinds:

1. τ permutes the set $X^{\pm 1}$ of generators (relabelling automorphisms).
2. For some fixed "multiplier" $a \in X^{\pm 1}$, τ sends each $x \in X^{\pm 1}$ to one of $x, xa, a^{-1}x$, or $a^{-1}xa$.

Note that the set of Whitehead automorphisms is larger than the set of Nielsen automorphisms, which already generates $\mathrm{Aut}(F_k)$. This larger set of automorphisms is needed to make the following algorithm work.

Whitehead's Algorithm. Given the pair (u, v), if some Whitehead automorphism reduces the length of u or v, apply that automorphism to obtain a new pair (u_1, v_1). Repeat until no Whitehead automorphism reduces the length of either member of the pair (u_t, v_t). The number of steps required to arrive at such a pair is clearly linear in $|u| + |v|$. If the lengths of u_t and v_t are not equal then there is no automorphism of F_k taking u to v. This is sometimes called the "easy part" of Whitehead's algorithm.

If $|u_t| = |v_t|$ and some automorphism takes u to v, then some sequence of Whitehead automorphisms takes u_t to v_t such that the length of all images remains equal to $|v_t|$.

The possible "problem" with the second step is that there are exponentially many words of a fixed length, and this step is sometimes called the "hard part" of Whitehead's algorithm. The algorithm is thus at worst single exponential time, but we do not know the exact time complexity of Whitehead's algorithm.

Definition 10. Call a Whitehead automorphism *nontrivial* if it is of the second kind (not a relabelling) and is not a conjugation. An element $w \in F_k$ is *strictly minimal* if it is cyclically reduced and its cyclically reduced image under every nontrivial Whitehead automorphism has length strictly greater than $|w|$.

The following theorem is from [6]:

Theorem 11. *Let C denote the set of cyclically reduced elements of F_k. The set of strictly minimal elements of C is strongly generic in C.*

It turns out that there is a very simple sufficient "counting criterion" for strict minimality. Namely, the number of all occurrences of one and two letter subwords should be close to their expected values.

Theorem 12 (Strict Minimality Criterion). *Let $0 < \epsilon < \frac{2k-3}{k(2k-1)(4k-3)}$. Suppose that w is a cyclically reduced word of length n such that*

(1) *For every letter $x \in \Sigma = X^{\pm 1}$, the number of occurrences of x in w divided by n is in the interval $\left(\frac{1}{2k} - \frac{\epsilon}{2}, \frac{1}{2k} + \frac{\epsilon}{2}\right)$;*
(2) *For every reduced pair of letters $p \in \Sigma^2$, the number of occurrences of p in w divided by n is in the interval $\left(\frac{1}{k(2k-1)} - \epsilon, \frac{1}{k(2k-1)} + \epsilon\right)$.*

Then w is strictly minimal.

So it is not necessary to apply any automorphisms in order to check that a word is strictly minimal! Technically, "Large Deviation Theory" supplies the proof.

The result shows that Whitehead's algorithm generically works in linear time. Experience with programming the algorithm had long showed that the algorithm usually runs quickly. Bilal Kahn proved that the algorithm is polynomial time in the case of two generators. Computer experiments show that positive random words are generically strictly minimal.

5 Isomorphism Rigidity of One-Relator Groups

It is not known if the Isomorphism Problem is solvable for one-relator groups. However, the problem is generically easy. In fact, we have a strong analog of Mostow rigidity for random one-relator groups. Fix the generating set $\{x_1, \ldots, x_k\}$. Two random one-relator groups on the given generators are isomorphic if and only if their Cayley graphs are isomorphic as labelled graphs, where the isomorphism is only allowed to permute the edge labels.

Isomorphism rigidity is most probably also true for random m-relator groups. One way to prove that is to positively settle the following conjecture.

Conjecture 13 (Stable Small Cancellation Conjecture). For any number $k > 1$ of generators and any $m > 1$, there is a generic set S of m-tuples such that for any automorphism α of the ambient free group F_k and any tuple $\langle w_1, \ldots, w_m \rangle \in S$, the set $\{\alpha(w_1), \ldots, \alpha(w_m)\}$ satisfies the small cancellation condition $C'(1/8)$.

6 Rewriting One-Relator Groups and Mapping Tori

The standard way to study a one-relator group is to rewrite the group as an HNN-extension of a one-relator group with shorter defining relator. This may require adding a root of a generator if no generator has exponent sum 0.

Consider the group $\langle x, y; xy^{-1}xy^{-1}xy^{-2} \rangle$, so $\sigma_x = 3, \sigma_y = -4$. Add a cube root to y. Thus substitute $x \to xy^4, y \to y^3$, giving

$$xy^4 y^{-3} xy^4 y^{-3} xy^4 y^{-6} = xyxyxy^{-2},$$

which we rewrite by subscripting occurrences of x by the exponent sum on y preceding the occurrence, giving $x_0 x_1 x_2 = 1$. Eliminating x_2 gives

$$G = \langle x_0, x_1, y; \ yx_0y^{-1} = x_1, \ yx_1y^{-1} = x_1^{-1}x_0^{-1} \rangle.$$

So G is the mapping torus of the automorphism which sends $x_0 \to x_1$, $x_1 \to x_1^{-1}x_0^{-1}$.

The interesting question is: How often does rewriting a two-generator one-relator group yield a mapping torus? Obtaining a mapping torus is *not* a generic property. This is proved by Nathan Dunfield and Dylan Thurston in [1].

Computer experiments with rewriting random words clearly confirm that the fraction of words whose rewritten form gives a mapping torus is between .9000 and .9200.

An interesting phenomenon is that when only *positive* words are considered the fraction of mapping tori goes up considerably, to around .99 in one experiment.

7   Translation Equivalence and Trace Identities

Results about this topic are taken from [4].

Definition 14. Let F be a finitely generated free group and let $u, v \in F$. We say that g and h are *translation equivalent*, written $u \equiv_t v$, if for every free and discrete action of F on an $\mathbb{R}$-tree, the translation lengths of u and v are equal.

Theorem 15. $u \equiv_t v$ *if and only if the cyclically reduced lengths of $\phi(u)$ and $\phi(v)$ are equal for every automorphism ϕ of F.*

Definition 16. If F is a finitely generated free group and $u, v \in F$ then u and v are *character equivalent* if for every representation $\rho: F \to SL(2, \mathbb{C})$ the equality $\mathrm{tr}(\rho(u)) = \mathrm{tr}(\rho(v))$ holds.

In the paper it is shown that character equivalence implies translation equivalence but the converse is false and thus translation equivalence is more general. Let $w(x, y) \in F(x, y)$ be a freely reduced word. By w^R we mean the word w written backwards, without inverting the letters. So $(xyxxy^{-1})^R = y^{-1}xxyx$.

Theorem 17. *Let F be a free group of rank $k \geq 2$ and let $w(x, y) \in F(x, y)$ be a freely reduced word. Then for any $u, v \in F$ we have*

$$w(u, v) \equiv_t w^R(u, v) \text{ in } F.$$

The fascinating subject of trace identities is discussed in detail in [2].

Question 18. Are there any nontrivial trace identities in $SL(3, \mathbb{C})$?

Computer experiments on random words using random representations have so far failed to find any.

8   Generic Stretching Factors for Free Group Automorphisms

The following result is from [3]:

Theorem 19. *Let $F = F_k(X)$ be the free group of rang $k \geq 2$ with basis X. If ϕ is any injective endomorphism of F, then there is a unique ration number $\lambda \geq 1$ such that, for all $\varepsilon > 0$,*

$$\left\{ w \in F : \left| \frac{|\phi(w)|}{|w|} - \lambda \right| < \varepsilon \right\}$$

is strongly generic in F. Furthermore, $\lambda = \lambda(\phi)$ is a computable rational number with

$$2k\lambda \in \mathbb{Z}\left[\frac{1}{2k-1}\right].$$

The theorem says that the images of long random words under ϕ are essentially stretched by the factor λ. The proof makes essential use of Kingman's Subadditive Ergodic Theorem.

9 Conclusion

The "philosophy" is exactly that expressed by Archimedes. Computer experiments can indeed discover what is true in certain situations. But even more, we see today that computer experiments are inextricably linked with genericity. If one is testing whether or not a property holds on random elements what one sees in a computer experiment is the "generic" behaviour of the property. If one is testing an algorithm on random inputs, what one sees is the generic behaviour of the algorithm. Such experiments have led to the realization that many very diverse properties hold generically and to the development of generic-case complexity for algorithms.

References

1. N. Dunfield, D. Thurston, A random tunnel-number one 3-manifold does not fiber over the circle. Geom. Topol. **10**, 2431–2499 (2006)
2. R.D. Horowitz, Characters of free groups represented in the two-dimensional special linear group. Commun. Pure Appl. Math. **25**, 635–649 (1972)
3. V. Kaiminovich, I. Kapovich, P. Schupp, The subadditive Ergodic theorem and generic stretching factors for free group automorphisms. Isr. J. Math. **157**, 1–46 (2007)
4. I. Kapovich, G. Levitt, P. Schupp, V. Shpilrain, Translation equivalence in free groups. Trans. Am. Math. Soc. **359**, 1527–1546 (2007)
5. I. Kapovich, I. Rivin, P. Schupp, V. Shpilrain, Densities in free groups and $\mathbb{Z}^k$, visible points and test elements. Trans. Am. Math. Soc. **359**, 1527–1546 (2007)
6. I. Kapovich, P. Schupp, V. Shpilrain, Generic properties of Whitehead's algorithm and isomorphism rigidity of one-relator groups. Pac. J. Math. **223**, 113–140 (2006)

Orbit Decidability, Applications and Variations

Enric Ventura

1 Orbit Decidability

In many areas of mathematics and in innumerable topics and situations, the notion of *transformation* plays an important role. If X is the set or collection of objects we are interested in, a transformation of X is usually understood to be just a map $\alpha \colon X \to X$. And whenever the context highlights a certain collection of "interesting" maps $A \subseteq \mathrm{Map}(X, X)$ (namely, endomorphisms or automorphisms of X if X is an algebraic structure, continuous maps or isometries of X if X is a topological or a geometric object, etc.), one naturally has the notion of orbit: the *A-orbit* of a point $x \in X$ is the set of all its A-images $xA = \{x\alpha \mid \alpha \in A\} \subseteq X$. In all these situations, there is a problem which is usually crucial when studying algorithmic aspects of many of the interesting problems one can formulate about the objects in X and how do they relate to each other under the transformations in A, namely orbit decidability.

Definition 1. Let X be a set, and let $A \subseteq \mathrm{Map}(X, X)$ be a set of transformations. We say that A is *orbit decidable* (*OD* for short) if there is an algorithm which, given $x, y \in X$, decides whether $x\alpha = y$ for some $\alpha \in A$. (Sometimes the algorithm is required to provide such an α, if it exists.)

There are lots of examples of very classical algorithmic problems which are of this kind. For example, the conjugacy problem of a group G is just the orbit decidability for the set of inner automorphisms $A = \mathrm{Inn}(G)$ (and recall that the word problem of G is a special subproblem). The classical Whitehead algorithm for the free group F_n is just a solution to the orbit decidability of the full automorphism group $A = \mathrm{Aut}(F_n)$, and all the variations of this problem (replacing elements to

E. Ventura (✉)
Departament de Matemàtica Aplicada III, Universitat Politècnica de Catalunya, Manresa, Spain
e-mail: enric.ventura@upc.edu

J. González-Meneses et al. (eds.), *Extended Abstracts Fall 2012*,
Trends in Mathematics 1, DOI 10.1007/978-3-319-05488-9_17,
© Springer International Publishing Switzerland 2014

conjugacy classes or subgroups, of tuples of them, etc.; replacing automorphisms to certain kind of automorphisms or endomorphisms, etc.; moving to other families of groups G or algebraic structures, etc.) are nothing else than other instances of orbit decidability.

A recent result by Bogopolski–Martino–Ventura [3] gave a renewed protagonism to the notion of orbit decidability. We first remind a couple of other concepts. The *twisted conjugacy problem* (*TCP*) for a group G consists of deciding, given $\alpha \in \mathrm{Aut}(G)$ and two elements $u, v \in G$, whether there exists $x \in G$ such that $(x\alpha)^{-1}ux = v$. Note that if α is the identity this is precisely the standard conjugacy problem (*CP*) for G; however, in general, it is a strictly stronger algorithmic problem (see [3, Corollary 4.9] for an example of a group with solvable *CP* but unsolvable *TCP*). On the other hand, for a short exact sequence of groups

$$1 \longrightarrow F \overset{\alpha}{\longrightarrow} G \overset{\beta}{\longrightarrow} H \longrightarrow 1,$$

and since $F\alpha$ is a normal subgroup of G, for every $g \in G$, the conjugation γ_g of G induces an automorphism of F, $\varphi_g \colon F \to F$, $x \mapsto g^{-1}xg$ (which does not necessarily belong to $\mathrm{Inn}(F)$). The set of all such automorphisms, $A_G = \{\varphi_g \mid g \in G\}$, is a subgroup of $\mathrm{Aut}(F)$ called the *action subgroup* of the given short exact sequence.

Theorem 2 (Bogopolski–Martino–Ventura [3]). *Let* $1 \longrightarrow F \overset{\alpha}{\longrightarrow} G \overset{\beta}{\longrightarrow} H \longrightarrow 1$ *be a short exact sequence of groups (given by finite presentations and the images of generators) such that*

(i) *F has solvable TCP,*
(ii) *H has solvable CP, and*
(iii) *For every $1 \neq h \in H$, the subgroup $\langle h \rangle$ has finite index in its centralizer $C_H(h)$, and there is an algorithm which computes a finite set of coset representatives, $z_{h,1}, \ldots, z_{h,t_h} \in H$ (i.e., $C_H(h) = \langle h \rangle z_{h,1} \sqcup \cdots \sqcup \langle h \rangle z_{h,t_h}$).*

Then G has solvable CP if and only if $A_G = \{\varphi_g \mid g \in G\} \leqslant \mathrm{Aut}(F)$ is OD.

Hypothesis (iii) is somehow restrictive, but at the same time satisfied by many groups: for example, free groups (where the centralizer of an element $1 \neq h$ is cyclic and generated by its maximal root) and it is not difficult to see that torsion-free hyperbolic groups also satisfy it; see [3, Subsection 4.2].

The correct way to think about this theorem is the following: it reduces the *CP* for a group G to the *TCP* plus a certain *OD* problem for a certain subgroup $H \leqslant G$. It is true that the *TCP* is harder than the standard *CP*, and the resulting *OD* problem is sometimes more technical than the original problem; but both of them take place *in the subgroup H* rather than in G. In all situations when H is a group significantly easier than G, Theorem 2 reduces the *CP* for G to two independent problems, maybe more technical but in an easier group H. Let us say it in a different way: for any group H where one knows how to solve the *TCP*, Theorem 2 gives a great tool to investigate the solvability/unsolvability of the *CP* in a vast family of extensions of H, by means of finding orbit decidable/orbit undecidable subgroups of $\mathrm{Aut}(H)$.

2 Applications

The idea behind Theorem 2 has proven to be quite fruitful, being the starting point of a collection of papers and preprints. The first one was [2], where Bogopolski–Martino–Maslakova–Ventura solved $TCP(F_n)$; combining this with Brinkmann's result that cyclic subgroups of $\mathrm{Aut}(F_n)$ are OD (see [5]), one immediately gets a solution to the CP for free-by-cyclic groups. (We remark that all these arguments made a crucial use of a result of Maslakova [10] on computability of the fixed subgroup of an automorphism of a free group, which is now under revision because of incorrectness of the original argument; see [1]. For an alternative solution to the CP for free-by-cyclic groups given by Bridson–Groves, see [4].)

In Theorem 2, we can take both F and H to be free groups. But a well known construction due to C. Miller (see [11]), provided examples of free-by-free groups with unsolvable CP. Hence, Theorem 2 tells us that $\mathrm{Aut}(F_n)$ must contain orbit undecidable subgroups $A \leqslant \mathrm{Aut}(F_n)$. This is not the case in rank 2 (every finitely generated subgroup of $\mathrm{Aut}(F_2)$ is OD; see [3, Proposition 6.13]), but they certainly do exist for higher rank, $n \geqslant 3$. A closer look to these negative examples revealed a general way to construct orbit undecidable subgroups inside $\mathrm{Aut}(G)$, as soon as $F_2 \times F_2$ embeds into it (see [3, Section 7]). This allowed to construct lots of new extensions of groups with unsolvable conjugacy problem. For example, since F_2 embeds in $GL_2(\mathbb{Z})$, $F_2 \times F_2$ embeds in $GL_4(\mathbb{Z})$ and one can deduce that $GL_4(\mathbb{Z}) = \mathrm{Aut}(\mathbb{Z}^4)$ contains orbit undecidable subgroups which, via Theorem 2, implies the existence of $\mathbb{Z}^n$-by-free groups with $n \geqslant 4$ and unsolvable CP (see [3, Proposition 7.5]). At this point it is worth mentioning that none of these arguments apply to the case of dimension 3 so, at the time of writing, it is an open problem whether there exists $\mathbb{Z}^3$-by-free groups with unsolvable CP (i.e., whether or not $GL_3(\mathbb{Z})$ contains orbit undecidable subgroups).

These last results were used by Sunic–Ventura in [13] to see that there exist automaton groups (i.e., subgroups of the automorphism group of a regular rooted tree, generated by finite self-similar sets) with unsolvable CP. In fact, in [13] and using techniques of Brunner and Sidki, it was proved that $\mathbb{Z}^d \rtimes \Gamma$ is an automaton group for every finitely generated $\Gamma \leqslant GL_d(\mathbb{Z})$. Then, by modifying the construction in [3] at the cost of increasing the dimension in 2 units, a finitely generated, orbit undecidable, free subgroup Γ of $GL_d(\mathbb{Z})$ was constructed, for $d \geqslant 6$. Using both results together with Theorem 2, one gets automaton groups with unsolvable CP (and additionally being [free-abelian]-by-free).

In the preprint [9], González-Meneses and Ventura consider the braid group B_n and solve $TCP(B_n)$. With a first superficial look, it may seem an easy problem because it is well known that $\mathrm{Out}(B_n) \simeq C_2$, with the non-trivial element represented by the automorphism $\alpha \colon B_n \to B_n$ which inverts all generators, $\sigma_i \mapsto \sigma_i^{-1}$. However, the conjugacy problem twisted by this α (namely solving the equation $(x\alpha)^{-1}ux = v$ for $x \in B_n$) becomes a quite delicate combinatorial problem about palindromic braids (see [9] for details). Furthermore, it is easy to see that every finitely generated subgroup $A \leqslant \mathrm{Aut}(B_n)$ is orbit decidable; hence, every extension of B_n by a torsion-free hyperbolic group H has solvable CP; see [9, Section 5].

A kind of opposite situation happens in Thompson's group F. Here, the automorphism group is quite big; but it is known that every automorphism of F can be realized as the conjugation by some element in $\widetilde{EP}_2$ (a certain discrete subgroup of Homeo([0, 1]) containing F). So, F has lots of automorphisms, but they all are structurally easy. This allowed Burillo–Matucci–Ventura to solve $TCP(F)$ in [6]. Since it is also proved there that $F_2 \times F_2$ does embed in Thompson's group F, one deduces the existence of Thompson-by-free groups with unsolvable CP.

A similar project is currently being carried over by Fernández-Alcober, Ventura and Zugadi for the family of Grigorchuk–Gupta–Sidki groups [8].

We encourage the (algorithmic oriented) reader to push the same idea further into his own area of expertise: choose your favorite group G, and try to solve $TCP(G)$. This will not be a very interesting result by itself (it is just a technical variation of $CP(G)$), but it will pave the way (via Theorem 2) to study the CP in a vast collection of extensions of G. You will have chances to prove results of the type "all G-by-[torsion-free hyperbolic] groups have solvable CP", or "there exists a G-by-free group with unsolvable CP".

3   Variations on Orbit Decidability

The definition of orbit decidability admits variations, pointing to deeper algorithmic problems. We present here one of these possible variations that we find interesting. It is not totally clear, by the moment, whether is it related to some algebraic problem, like standard orbit decidability is related to the CP via Theorem 2. Even if it is not, the problems it provides are interesting enough by themselves.

Definition 3. Let G be a group, and $A \leqslant \text{Aut}(G)$. We say that A is $(m\text{-})subgroup$ orbit decidable, $(m\text{-})SOD$ for short, if there is an algorithm which, given $g, h_1, \ldots, h_m \in G$, decides whether $g\alpha \in H = \langle h_1, \ldots, h_m \rangle \leqslant G$ for some $\alpha \in A$.

Since in F_n, as well as in $\mathbb{Z}^n$, roots of elements are well-defined and must be preserved by automorphisms (i.e., $x\alpha = y$ implies $\hat{x}\alpha = \hat{y}$), it is easy to see that, for every A, solvability of $OD(A)$ implies solvability of $1\text{-}SOD(A)$. However, $m\text{-}SOD(A)$ for $m \geqslant 2$ looks like a much more complicated problem, even over the free abelian group.

Over the free group F_n, two special instances of this problem are solved in the literature. Silva–Weil solved in [12] the problem $SOD(\text{Aut}(F_2))$: given an element x and a subgroup H of the rank two free group F_2, one can algorithmically decide whether $x\alpha \in H$ for some $\alpha \in \text{Aut}(F_2)$. And Clifford–Goldstein [7] gave an algorithm solving the particular case of $SOD(\text{Aut}(F_n))$ where the given input x is a primitive element: there is an algorithm deciding whether a given subgroup $H \leqslant F_n$

contains a primitive element of F_n. The rest of the problem $SOD(\text{Aut}(F_n))$ remains open, and nothing is know for other subgroups $A \leqslant \text{Aut}(F_n)$.

Over the free abelian group $\mathbb{Z}^n$, $SOD(GL_n(\mathbb{Z}))$ is an exercise (just a matter of gcd's of the entries of the involved vectors). But, for a fixed given matrix $A \in GL_n(\mathbb{Z})$, the problem $SOD(\langle A \rangle)$ is much more interesting: after projectivizing $\mathbb{Z}^n$, the automorphism $A : \mathbb{Z}^n \to \mathbb{Z}^n$ induces a map $\varphi : \mathbb{P}^{n-1}(\mathbb{Z}) \to \mathbb{P}^{n-1}(\mathbb{Z})$, and $SOD(\langle A \rangle)$ becomes the problem of deciding whether a given orbit of φ intersects a given (projective) linear variety in $\mathbb{P}^{n-1}(\mathbb{Z})$ (for $n = 2$, this problem becomes a nice exercise in linear algebra, involving the eigenvalues of A).

Acknowledgements The author thanks the CRM for its hospitality during the research programme on Automorphisms of Free Groups. He also acknowledges partial support from the Spanish Government through grant number MTM2011-25955.

References

1. O. Bogopolski, O. Maslakova, A basis of the fixed point subgroup of an automorphism of a free group (preprint). http://es.arxiv.org/PS_cache/arxiv/pdf/1204/1204.6728v6.pdf
2. O. Bogopolski, A. Martino, O. Maslakova, E. Ventura, Free-by-cyclic groups have solvable conjugacy problem. Bull. Lond. Math. Soc. **38**(5), 787–794 (2006)
3. O. Bogopolski, A. Martino, E. Ventura, Orbit decidability and the conjugacy problem for some extensions of groups. Trans. Am. Math. Soc. **362**, 2003–2036 (2010)
4. M. Bridson, D. Groves, *The Quadratic Isoperimetric Inequality for Mapping Tori of Free Group Automorphisms*. Memoirs of the American Mathematical Society, vol. 203, no. 955 (American Mathematical Society, Providence, 2010)
5. P. Brinkmann, Detecting automorphic orbits in free groups. J. Algebra **324**, 1083–1097 (2010)
6. J. Burillo, F. Matucci, E. Ventura, The conjugacy problem for extensions of Thompson's group (preprint). http://es.arxiv.org/pdf/1307.6750
7. A. Clifford, R. Goldstein, Subgroups of free groups and primitive elements. J. Group Theory **13**(4), 601–611 (2010)
8. G. Fernández-Alcober, E. Ventura, A. Zugadi, On the twisted conjugacy problem for GGS-groups (work in progress)
9. J. González-Meneses, E. Ventura, Twisted conjugacy in the braid group to appear at Israel J. Math., http://download.springer.com/static/pdf/305/art%253A10.1007%252Fs11856-014-0032-4.pdf?auth66=1404553037_a859d544001f67d55b94de03e253fd73&ext=.pdf
10. O. Maslakova, Fixed point subgroup of an automorphism of a free group. Algebra Log. **42**(4), 422–472 (2003) (in Russian)
11. C.F. Miller III, *On Group-Theoretic Decision Problems and Their Classification*. Annals of Mathematics Studies, vol. 68 (Princeton University Press, Princeton, 1971)
12. P. Silva, P. Weil, Automorphic orbits in free groups: words versus subgroups. Int. J. Algebra Comput. **20**(4), 561–590 (2010)
13. Z. Sunic, E. Ventura, The conjugacy problem in automaton groups is not solvable. J. Algebra **364**, 148–154 (2012)

Printed in the United States
By Bookmasters